Edward Powers

War and the Weather

Edward Powers

War and the Weather

ISBN/EAN: 9783337340490

Printed in Europe, USA, Canada, Australia, Japan

Cover: Foto ©ninafisch / pixelio.de

More available books at **www.hansebooks.com**

WAR AND THE WEATHER,

OR, THE

ARTIFICIAL PRODUCTION OF RAIN.

By EDWARD POWERS, C. E.

CHICAGO:
S. C. GRIGGS & COMPANY,
117 & 119 STATE STREET.
1871.

THE

ARTIFICIAL PRODUCTION OF RAIN.

The idea that rain can be produced by human agency, though sufficiently startling, is not one which, in this age of progress, ought to be considered as impossible of practical realization. It is an opinion of comparatively recent origin, and is one which cannot be regarded as belonging, in any degree, to a certain class of notions which prevail among the unthinking, and which, being based neither on reason nor observed facts, are respectable, if at all, only for their antiquity; but, on the contrary, it is one which is confined principally to those who are accustomed to draw conclusions only from adequate premises, and whose belief in the matter referred to has generally been founded on facts which have come under their own observation. When numerous observers, each independently of the others, arrive at an identical conclusion, in reasoning from facts which they have separately noticed in widely different fields, such conclusion is certainly worthy of respect, and may be assumed to contain the elements of truth. Of this nature is the idea under consideration — the belief that rain has been, and can be, brought on by heavy discharges of artillery.

In collecting some of the facts bearing on this question and submitting them to the public, the object of the writer has been to awaken a more general interest in the subject, in the hope that Congress may be induced to cause some experiments to be made for the purpose of developing the natural principle that seems to be involved, and determining if it cannot be made of practical use to the country. If it should be conceded — as it must be from the evidence that will be presented—that battles have produced changes in the weather, it would seem to be an eminently proper subject for legislative action to provide for an investigation of the conditions under which these changes can be made. If lightning and thunder and rain have been brought on by the agency of man, when bloodshed and slaughter only were intended, this surely can be done without these latter concomitants. And when we consider the grand results that would flow from an assured power and well defined method of causing rain to fall at will — the mighty step that would thereby be made by man towards the complete control over nature to which he aspires— the bare possibility that such a power, heretofore considered as a prerogative of the Deity alone, is within his reach, ought to be sufficient to lead to an earnest inquiry into the truth of the matter, and to an investigation as to the most economical and effective means of applying it, if it should be found to exist. That there is the strongest reason for believing that this achievement is possible, I have the means of showing; but to verify the truth of the theory by which such power is claimed, and to determine its limits and conditions, can only be done by a well regulated series of experiments with powder and cannon and other appliances. Such experiments, when made, as eventually they surely will be, should be made at the public expense; for it is the public who would be benefited in the event of their success. The art of regulating the weather to some extent, if such an art should ever be acquired, is not one on which a patent could ever be obtained, nor would the business be one in which a monopoly could ever be exercised by an individual. The agricultural class, it is true, would be the one which would be the most directly benefited by it, but the prosperity of this class, as a general rule, leads to the prosperity of all the others.

Before submitting the evidence by which I propose to show the connection between artillery firing and rains, or endeavoring to present a reasonable theory for the assumed direct relation between the two, it may be well to offer a few remarks in regard to the commonly accepted theory of rains in general. The air, as is well understood, is the great reservoir in which is collected and stored up the water from which all storms are formed. Extending around the earth to the height of forty or fifty miles, it is capable of holding in suspension a vast amount of this fluid, which it receives from evaporation from the ocean, from lakes, rivers, pools, and from all portions of the earth's surface where any moisture is present. The water, when so evaporated, passes into the air in the form of a transparent and perfectly invisible vapor, and the warmer the air, the greater the amount of this vapor it is capable of absorbing. Rain is formed by the condensation of this vapor and its precipitation to the earth; a partial condensation first forming clouds, and rendering the vapor visible. This condensation is supposed to be caused by the cooling of the air in which it occurs, whereby the amount of vapor which it is capable of holding is lessened. Thus a warm current of air saturated with moisture, meets a cooler current, and the cold of the latter condenses a portion of the aqueous vapor contained in the former, and clouds and rain are the result.

The theory of which the foregoing is an outline, is no doubt correct in its main features, but it stops short of a full explanation of the production of rain. It is founded on the known principle that two bodies of air of different temperatures, cannot hold as much aqueous vapor, after they are mixed together, as they were capable of holding while separate; and from this it shows how a thin cloud may be formed, but it fails to show how the particles of this cloud are collected into drops of rain. In recognizing cold, or the absence of a certain degree of heat, as the only cause of the condensation of aqueous vapor into clouds and rain, it is unsatisfactory. Without going into any extended argument to show that there must be some other agent as well as the heat at work to produce this result, the objection will simply be made here that the theory referred to requires that immense quantities of air of different temperatures should be mixed for the production of a very small amount of rain; that the two bodies of air so mixed should be fully saturated with moisture, or nearly so—otherwise the resulting mixture would be capable of containing all, or nearly all, the moisture previously contained in both; and finally it teaches that this air, instead of being deprived of the greater portion of its moisture by the formation of rain, is left as full of moisture as it can possibly hold at the temperature * which results when the mixing process takes place.

These considerations, and especially the latter, should impel us to look for some further cause for the condensation of aqueous vapor in the production of rain. This cause may be some form of electrical action. Electricity no doubt plays an important part in all storms, though it may not always manifest itself to the senses; and in a thunder storm it is as reasonable to suppose that the lightning, (or the electricity of which it is formed), is an agent in causing the rain, as that the rain

* Rain is generally produced by the rapid union of two or more volumes of humid air differing considerably in temperature; the several portions, when mingled, being incapable of absorbing the same amount of moisture that each would retain if they had not united. If the excess is great it falls as rain; if it is of slight amount it appears as cloud. The production of rain is the result of the law that the capacity of the air for moisture decreases in a higher ratio than the temperature. (*Silliman's Principles of Physics or Natural Philosophy*, page 656.)

causes the lightning; or that the process that produces the one, produces also the other.

But dismissing for the present the subject of theories, let us proceed to facts—facts, but few of which, perhaps, would be of great significance if they stood alone and unsupported by the others; but which, taken in the aggregate, furnish the strongest evidence that heavy artillery firing has an influence on the weather and tends to bring rain.

Let it be premised that this statement of facts, however, is far from being a complete one; and lacks much of exhibiting all the evidence that could be found in support of the above proposition. That there were a great number of minor engagements during our late war, not included in this list, that were followed by an early fall of rain, is shown by the general scope of the testimony which I shall append upon the subject. I have drawn many facts from the records of the Navy Department as contained in the Log Books preserved in the Office of the Bureau of Navigation, and probably many more could be obtained from the same source; but though furnished, through the kindness and courtesy of the Bureau, with every facility in the examination of these books, but a very limited time could be given to the work, and it has been necessarily hasty and far from complete. If the facts, however, which I shall present, are insufficient to convince, it would be in vain to hope to do so with a greater number.

The first instance that I shall mention of rain as a direct result of a battle, occurred at

THE BATTLE OF BUENA VISTA, MEXICO,

fought 22nd and 23rd of February, 1847. This was in the dry season in that country; there had been no rain for several months before the battle, and there was none for several months after. Three showers, however, followed the first day's engagement,[1] two of which are specially remarkable. On the 23rd, about one or two hours after the severe cannonading that took place between 8 and 10 A. M., there was a most violent rain fall for some ten or fifteen minutes. Again, in the afternoon, at about the same interval, after another spell of heavy cannonading, another violent shower of rain fell. The fact before stated, that no rain had fallen for months before the battle, and that none fell for months after at that place, is almost proof positive, not only that the cannonading caused the rain, but that the cannonading will bring rain at a time when the atmospheric conditions are in the highest degree unfavorable to the production of that phenomenon through the ordinary operations of nature.

THE BATTLE OF PALO ALTO, MEXICO,

8th May, 1846; also fought in the dry season, was also followed by rain,[2] but the particulars cannot be given.

THE BATTLE OR SIEGE OF MONTEREY, MEXICO,

was fought September 21 to 23, 1846, which, like

[1] See appended documents, Nos. 7 and 27.
[2] See Doc. No. 30.

the dates before mentioned was in the dry season. The morning of the 21st was bright and beautiful, but " soon after the storming of the two forts, Federacion and Soldado, a violent storm came up," and to its " unbroken peltings" General Worth and the 2nd Division were exposed during the night.[3] A similar phenomenon occurred also on the 23rd. The morning was bright and fair, with no indications of rain ; during the day there was heavy cannonading, and " the evening and night closed in with heavy rain."[4]

THE BATTLE OF CONTRERAS, MEXICO,

was fought August 19th, 1847, and at a season when rains were very unfrequent. At Puebla during the months of June and July, and perhaps the beginning of August, there had been heavy falls of rain *every afternoon*, the skies clearing before sunset, and the atmosphere being remarkably clear until the next afternoon. Our army commenced its march from Puebla on the 7th, and from that time until the 16th, the days were generally, if not always, clear, bright, and beautiful.[5] On the 16th the movement commenced at Chalco, and on the next two days there was some rain, but it was not heavy. The morning of the 19th was bright and clear,[6] and such was also the afternoon[7] at the usual hour for rains.

But on this day " the battle raged furiously, and for more than three hours the entire force was

---

[3] See Doc. No. 27, D. 30.   [4] See Doc. No. 28.   [5] See No. 26.
[6] See Doc. No. 28.   [7] See No. 26.

under a heavy fire." " Night at length put an end to the conflict"; and " a cold rain soon afterwards began to fall in torrents."[8]

THE BATTLE OF CHURUBUSCO

was fought the next day, which was bright and clear. The day after it rained heavily.[9]

THE BATTLE OF MOLINO DEL REY, MEXICO,

fought September 8, 1847, was also followed in the afternoon and evening by a hard rain.[10]

THE BATTLE OF CHEPULTEPEC, MEXICO,

fought September 13, 1847, was also followed by rain; and whatever doubt may be entertained as to the significance of the facts of rain following the battles of Contreras and Churubusco, on account of their nearness to the wet season, it cannot be doubted that at this time the dry season had fully set in. The day of the battle was followed by a dark and cloudy night,[11] with rain in the early morning of the day following. The Mexican historian says: " The morning of the 14th was as gloomy and sad as the destiny of the Republic. There was a mist so thick that objects could not be seen at a few steps distance. Soon after, a light shower began to fall which soaked the soldiers,"[12] etc. Later in the morning the weather became clear.

During the late war of the rebellion, the occur-

---

[9] See No. 26. [8 & 10] See No. 27.
[11] See Appended Document, No. 27. [12] See No. 30.

rence of the phenomenon under discussion was frequent. The

BATTLE OF BIG BETHEL

may be mentioned as an early instance. This battle, fought in Eastern Virginia on the 10th of June, 1861, was soon followed by a copious rain.[13]

Incessant rains attended General McClellan's

CAMPAIGN IN WESTERN VIRGINIA,

in July, 1861. It has been published that his troops "had four separate engagements on four days, and before the close of each, violent rains fell."[14] The

BATTLE OF RICH MOUNTAIN,

fought July 10, was one of these, and was followed by one or two rainy days.[15]

The following engagements, which took place in that section of the country later in the same year, were also each followed quickly by rain; viz.:

BATTLE OF CARNIFAX FERRY,[16]

fought August 10, 1861.

BATTLE OF CHEAT MOUNTAIN,[17]

fought 13th and 14th September, 1861.

BATTLE OF GREEN BRIER,[18]

fought October 3, 1861.

[13] See No. 39. [14] See page 72. [15] See Doc. No. 5. [16] See No. 17. [17] and [18] See No. 38.

BATTLE OF ALLEGHANY SUMMIT,[19]

fought December 12, 1861.

None of these are classed as great battles, but the firing was, apparently, sufficient to bring rain. At the first great battle of the war the resulting phenomenon was similar, but more intensified. This, the

FIRST BATTLE OF BULL RUN,

was fought on the 21st of July, 1861. The day of the battle was bright and clear all through, but the next day was one of drenching rain. The storm commenced about six o'clock in the morning and continued all day and through the following night; the rain, during the afternoon and night especially, falling in torrents.[20]

As early in the war as the

SIEGE OF LEXINGTON, MISSOURI,

which ended on the 20th of September, 1861, in the surrender of Colonel Mulligan to the Confederates, the fact that heavy artillery firing was usually followed by rain, had already been noticed in the West. On the 17th the beleagured garrison were cut off from the river, and thus deprived of water; but to encourage the soldiers to hold out as long as possible for the arrival of the expected reinforcements, it was represented to them, by their officers, that the cannonading would surely bring

---

[19] See No. 38.   [20] See Appended Documents, Nos. 1, 3, 20 and 26.

rain to quench their thirst. And this prediction was fulfilled; though, unfortunately, they had no way to catch the water which their firing had drawn from the skies, except by spreading their blankets to the shower, and then wringing them out.[21]

In the South, as well as in the East and West, rain followed heavy cannonading. An

ENGAGEMENT NEAR FORT PICKENS, FLORIDA,

was an early instance. Flag Officer William W. McKean, commanding Gulf Blockading Squadron, in a report to the Secretary of the Navy, dated November 25, 1861, thus mentions the circumstance. He says: "Sir — I have the honor to inform you that on the 22nd instant, a combined attack was made upon the rebels at this place by Colonel Brown, of Fort Pickens, and the United States ships Niagara and Richmond under my command. * * * At ten o'clock, at the firing of the first gun from the Fort (the signal agreed upon), the Niagara stood in, followed by the Richmond, and both ships came to anchor. * * * We immediately opened fire. * * * At six P. M. a sudden squall came up from the Northward and Westward, the wind blowing very fresh, with heavy rain," etc.[22]

---

[21] See Doc. No. 4. See also Greeley's History of the American Conflict, Vol. I., page 588.

[22] See Documents accompanying Report of Secretary of Navy, of December 1, 1862.

In the middle portions of the country also, as well as in the East, West and South, the phenomenon referred to was, early in the war, exhibited. The

BATTLE OF LOGAN'S CROSS ROADS

is an instance. The author of the American Conflict, in speaking of the pursuit of the Confederates after this battle, says, "It rained as usual,"[23] a remark which is understood to recognize a truth which it is the object of this treatise to bring forward — and which receives still more pointed notice on a subsequent page of that work.[24]

THE BATTLE OF FORT DONELSON,

which all will remember as one of the first great victories of the war for the Union, affords another instance of the kind under consideration. The siege commenced on the 13th of February, 1862, which was a clear, bright day, as was also the next. The artillery firing commenced on the 14th, by a desperate fight of an hour's duration between Commodore Foote's gunboats and the batteries of the fort; the gunboats finally retiring, badly crippled. The next day the battle was renewed by the land forces, and ended in a storm of snow, which in turn was followed by one of rain.[25] The weather this day changed to cold — a change which, it is presumed, would have occurred if there had been no battle; but the effect of the cold was to turn in

[23] Vol. II., page 43.   [24] Vol. II., page 392.   [25] See Doc. No. 3.

part into snow and sleet the storm which the cannonading brought, and which would otherwise have probably been one wholly of rain.

THE BATTLE OF PEA RIDGE, ARKANSAS,

fought March 7 and 8, 1862, was followed on the morning of the 9th by a hard rain.[26]

All the important operations of the expedition which was sent under General Burnside and Commodore (now Rear Admiral) L. M. Goldsborough against the Confederate strongholds in North Carolina, were each quickly followed by rain. The first of these was the attack upon and capture of

ROANOKE ISLAND,

on the 7th and 8th of February, 1862. Commodore Goldsborough, in his official report, in speaking of the weather at daylight in the morning of the 7th, says " The sky gave evident signs of a clear day." In the course of the forenoon his gunboats commenced an attack on the rebel batteries and gunboats, which was continued through the day. In the night it commenced to rain,[27] and the next day was rainy throughout. On the second day of the fight, the engagement was renewed by the fleet, while the land forces assaulted and carried the works in the rear. The rain, which accompanied

---

[26] See Doc. No. 34.

[27] and [28] Log of the " Stars and Stripes." See also Am. Conflict, Vol. II., p. 75, and Doc. No. 35 hereto appended.

and followed this day's action, continued until noon of the day following.[28]

The next important movement of this expedition was against

NEWBERN, NORTH CAROLINA.

The city was taken on the 14th of March, 1862, but there was much heavy firing on the 13th by the gunboats, and in the night there was a pouring rain.[29] No rain fell on the day of the assault and capture of the enemy's works, but the day after was very rainy.[30]

The next and last important operation of the expedition above referred to was the

CAPTURE OF FORT MACON.

Fire was opened on this work from General Burnside's siege guns, on the 25th of April, 1862, at about six o'clock in the morning, and was continued until late in the afternoon; four of Commodore Gouldsbourgh's vessels also taking part in the action. The sky that morning was clear, and so remained until about six o'clock in the afternoon. At that hour it became overspread with clouds, and the next afternoon it rained,[31] the rain falling heavily for three hours.

---

[29] "Am. Conflict," Vol. II. p. 77.
[30] Log of the U. S. Steamer "Delaware."
[31] Log of U. S. Steamer "Daylight."

THE NAVAL ACTION IN HAMPTON ROADS,

in which the U. S. Ships Congress and Cumberland were destroyed by the Merrimack and other Confederate vessels, furnishes another instance of rain following the discharge of artillery. The fight took place on the 8th of March, 1862, a clear, cool day. The next day—the one on which the contest happened between the Merrimack and Monitor—there were four hours of drizzling rain.[32]

On the Mississippi River, scarcely an action of any moment occurred that was not followed immediately by rain. The engagement which resulted in

THE CAPTURE OF NEW MADRID, MISSOURI,

was a marked instance. The fight took place on the 13th of March, 1862; a heavy cannonade was kept up from both sides through the day, and a violent thunder storm raged through most of the night.[33]

AT ISLAND NO. TEN,

several instances of the kind occurred. The first was at the general attack that was made on the batteries of the island by Commodore Foote's flotilla. This attack was made on the 17th of March, 1862; and during the next day, also, the mortar vessels continued to throw shells into the rebel works. The weather on the first day was clear, and on the second it was the same until six P. M. At that

---

[32] Log of the U. S. Steamer "Roanoke." See also Doc. No. 39.
[33] Am. Conflict, Vol. II., p. 55.

hour the sky became overcast, and thunder showers followed in the latter part of the night.[34]

ANOTHER INSTANCE OF A SIMILAR KIND OCCURRED AT THE SAME PLACE,

a short time after. Under date of April 3, 1862, eight to twelve A. M., the Log of the "Benton" says: "Clear and calm and very warm. The Benton, Cincinnati and Carondelet have taken position along the Missouri shore and opened fire on the floating battery and Island. The mortars are also actively engaged." The weather which followed this engagement is thus stated, under date of April 4. From four to eight A. M., " Clear weather until six o'clock, then clouded up and threatened rain." From eight to twelve, " Fresh breeze from E. S. E., attended with much rain." From twelve to four P. M., " Passing showers from southward and eastward until two o'clock; two o'clock till four clear, with moderate breezes from S. W. S."

This rain, which took place on the fourth, we may suppose to have been brought on by the action of the third. But there was also an

ACTION ON THE MORNING OF THE FOURTH,

which, apparently, produced another rain, and one more violent than the first. The nature of this engagement is thus explained in a dispatch from Flag Officer Foote to the Secretary of the Navy, dated April 4, 1862. He says: " This morning the

[34] Log of the "Benton."

Benton, Cincinnati and Pittsburg, with three mortar boats, opened and continued for more than an hour, a fire on the rebels' floating battery at Island No. 10. * * * The shells were thrown from the flotilla into different forts of the island, and into the rebel batteries lining the Tennessee shore."

Continuing to quote from the Log Book of the Benton for April 4, 1862; the weather, a few hours after this action, and after the other shower had fully cleared away, is thus described: From four to six P. M., " Wind South"; from six to eight P. M., " Fresh breeze from S. and cloudy"; from eight to twelve P. M., " Variable winds and heavy showers of rain, accompanied with very vivid and constant lightning and some thunder." It was in this thunder storm that the gunboat Carondelet ran the rebel batteries.

There was still

ANOTHER RAIN FOLLOWING HEAVY CANNONADING AT ISLAND NO. TEN.

The firing, as referred to in one of the dispatches of Commodore Foote, dated April 8, 1862, occurred on the seventh. General Pope is spoken of as having crossed the river that day under a heavy fire, and reference is made to the reduction of a fort by two gunboats. The rain occurred on the morning of the eighth, and is mentioned in another dispatch of Commodore Foote of that date, as a heavy thunder storm.[35]

[35] See Documents accompanying Report of the Secretary of the Navy, of December 1, 1862.

As the Island surrendered on the eighth, I have no further thunder storms to record as following cannonading at that point.

At the

BOMBARDMENT OF FORTS JACKSON AND ST. PHILIP,

on the Mississippi River, below New Orleans, commencing April 18, 1862, two days of rather slow firing, in dry weather, by Farragut's fleet, was followed, on the morning of the third, by some four hours of rain.[36]

THE GRAND ATTACK UPON AND PASSAGE OF THE FORTS,

and destruction of the rebel fleet on the twenty-fourth, was followed by a terrific thunder storm, lasting about five hours. The attack commenced between three and four o'clock in the morning, under a sky which remained cloudless until four P. M.; and the rain commenced between eleven and twelve A. M. next day.[37] This furious storm was raging when the fleet reached the city of New Orleans.

A thunder storm also followed the

BOMBARDMENT AND PASSAGE OF THE VICKSBURG BATTERIES

by some vessels of Farragut's fleet, and of the mortar flotilla on the morning of June 28, 1862. The attack was made at three o'clock in the morning,

---

[36 and 37] Log of the Hartford. See also Am. Conflict, Vol. II., p. 94. Also appended Doc. No. 41.

with some thirteen or fourteen vessels, and while they were steaming past the city the firing was rapid and heavy. The weather at the time was clear, with the exception of some detached clouds, and so remained until midnight of that day. Through the remainder of the night clouds and lightning were seen, and the morning brought several hours of weather "squally, with rain and heavy thunder and lightning."[38]

Again, on the morning of June 30,

THE MORTAR VESSELS ENGAGED THE VICKSBURG BATTERIES.

The next morning there was a terrific thunder storm, lasting about two hours. The amount and intensity of the lightning, and the violence of the rain in this storm, were extraordinary.[39]

THE NAVAL ENGAGEMENT NEAR VICKSBURG,

on the morning of July 15, 1862, was also followed by a storm. An expedition had started up the Yazoo River, that morning, to destroy the rebel ram "Arkansas," when it met that vessel coming down. A severe fight ensued, in which two of our vessels were disabled; after which the Arkansas escaped into the Mississippi, and took refuge under the Vicksburg batteries; from which unsuccessful attempts were made that day to cut her out; the last being made in the evening. The weather was clear at the time this fighting commenced, as it had

[38 and 39] Log Book of the U. S. Steamship Hartford.

been for nearly two weeks previous; but between four and six in the afternoon there was a shower; and about midnight a heavy rain commenced which lasted, with varying intensity, until four o'clock in the afternoon of the following day.[40] After this there was another spell of dry weather, broken by

ANOTHER ENGAGEMENT AT VICKSBURG,

occurring on the 22nd of July, 1862. On this day the "Benton," "Louisville" and "Cincinnati" attacked the upper batteries, while the "Essex" and ram "Queen of the West" went down and attacked the rebel ram "Arkansas" in her place at the levee. The action commenced at 4.30 A.M., the weather being at the time clear and calm. Soon after the action a light variable wind sprang up. In the afternoon the sky became overclouded, with light west wind. From four to six P.M. clear and calm again. In the evening it again became overcast, with light wind first from south then from west, and after midnight variable. From noon to four P.M. fresh southwest wind with rain.[41]

Tremendous rains fell during the night of each day of

THE BATTLE OF PITTSBURG LANDING, OR SHILOH,

Tennessee, fought on the 6th and 7th of April, 1862. The morning of the sixth was clear and beautiful, with no indications of a storm, but the day's terrific battle was followed by a night of

[40] Log of the "Benton." [41] Log of the "Benton."

drenching rain.[42] The battle of the next day was also succeeded in the night by a fearful storm, which, in this case, consisted of rain, hail and sleet. "An impressed New Yorker," in writing of the retreat of the Confederate army on this terrible night, says: "And to add to the horrors of the scene, the elements of heaven marshalled their forces — a fitting accompaniment of the tempest of human desolation and passion which was raging. A cold drizzling rain commenced about nightfall, and soon came harder and faster, then turned to pitiless, blinding hail. This storm raged with unrelenting violence for three hours. I passed long wagon trains filled with wounded and dying soldiers, without even a blanket to shield them from the driving sleet and hail, which fell in stones as large as partridge eggs until it lay on the ground two inches deep."[49]

In the list of military conflicts followed by rain, is also to be placed the

BATTLE OF BULL PASTURE MOUNTAIN, OR M'DOWELL.

This battle was fought in Western Virginia, on the 8th of May, 1862; and the circumstances connected with the rain were such as to aid in confirming an officer, who was present, in his belief that rain was a direct result of battle.[50]

[42] See appended Documents Nos. 3, 14, 16 and 28.
[49] See Note on page 60, Vol. II. of "American Conflict."
[50] See Doc. No. 38.

In this list is also to be placed General Banks'

BATTLE OF WINCHESTER, VIRGINIA.

Banks' retreat from the Shenandoah Valley was made on the 24th and 25th of May, 1862; and the battle occurred on the morning of the 25th, which was a dry hot day. The weather which followed is thus referred to in history, in connection with the movement which was immediately made by General Fremont, with a view to intercept Jackson on his return from his pursuit of Banks; viz., "Through constant rains and over mountain roads that could be made barely passable, he crossed the Alleghanies and descended into the Valley."[51]

AFTER THE BATTLE OF CROSS KEYS,

fought in the Shenandoah Valley between Generals Fremont and Jackson, on the 8th day of June, 1862, it again rained in that section of the country,[52] and on the night of the ninth the rain had extended to the Southeastern part of the State.[53] The battle of Port Republic was fought by the same forces on the ninth, and again on the night of the tenth rain appeared in Southeastern Virginia.[54]

The history of

GENERAL M'CLELLAN'S CAMPAIGN ON THE PENINSULA,

from the investment of Yorktown to the sanguinary battle of Malvern Hill, presents a continued

[51] American Conflict, Vol. II., p. 137.
[52] See Documents 5 and 38.
[53, 54] See Document No. 20.

succession of battles and rains. His first advance was commenced on the fourth of April, 1862, and was stopped on the night of the sixth, by the fire of rebel batteries; and the cannonading that then and soon after ensued was followed on the seventh, eighth and ninth, at the point of observation of the U. S. Steamer Wachusett, with more or less rain each day. In tracing the subsequent rains, in connection with the history of the time, great assistance is rendered by some extracts which have been kindly furnished from the journal of Major General Heintzelman, who commanded an army corps in the campaign.

Commencing with the operations immediately preceding the

CAPTURE OF YORKTOWN,

we find that on the second of May, 1862, "some five hundred shot and shells were fired by the rebels." The next day it "threatened rain, but turned clear and pleasant."[55]

On the night of the third, "the rebels were very busy until after midnight firing (artillery)." On this night they abandoned their works at Yorktown, and being pursued the next day (May 4), there ensued, at

FORT MAGRUDER,

in front of Williamsburg, a sharp cannonade.[56]

[55] See Document No. 20.
[56] American Conflict, Vol. II., p. 122.

During the following night a heavy rain set in.[57] The next day was fought the

BATTLE OF WILLIAMSBURG.

It rained through the day and into the night following.[58]

THE FIGHTING ON THE CHICKAHOMINY,

between the advance of the army and the rebels, commenced on the twenty-fourth of May. On the twenty-seventh we read of "pouring rains."[59] At this date occurred the

BATTLE OF HANOVER COURT HOUSE,

and on the thirtieth there was a heavy thunder storm, the rain falling in torrents.[60]

On the afternoon of May 31, and morning of June 1, was fought the great

BATTLE OF FAIR OAKS, OR SEVEN PINES.

On the morning of the second it began to rain; during the night of that day it rained heavily; and on the night of the third and morning of the fourth the very flood-gates of heaven seemed to be opened. By the fearful rains which followed this battle, the surrounding country was flooded, and movements on the part of either army rendered, for a time, almost impossible.[61]

---

[57] See Document No. 20.
[58] See Document No. 20.
[59] "American Conflict," Vol. II., p. 141.
[60] See Document No. 20
[61] See Doc. No. 20. Also Gen. McClellan's official dispatches, etc.

The weather, after this rain, remained unsettled for some days—but without attempting to show a connection between this fact and the firing that occurred in the meantime between the two armies still facing each other on the Chickahominy, I will pass over a period of about two weeks to notice some

GUNBOAT FIRING ON JAMES RIVER.

In the journal of Major General Heintzelman it is recorded, under date of June 17, 1862, " The gunboats were firing nearly two hours to-day," and under date of the eighteenth, " Since dark a heavy wind and rain."[62]

ARTILLERY FIRING IN FRONT OF HOOKER.

Under date of June 21, we read : " Suddenly a brisk fire of musketry ran along Hooker's front, followed by artillery ; " and under date of the twenty-second, " We have had a few drops of rain since dark." On the twenty-third, also, there were " Showers of rain with a little thunder."[63]

MORE ARTILLERY FIRING.

Under date of the twenty-fourth it is recorded : " At dawn heavy musketry commenced, soon followed by artillery ;" and, " Had another heavy rain a little before night."[64]

Let me pause here a moment to remark, what indeed must be obvious, that neither artillery firing

---

[62, 63 and 64] See Document No. 20.

nor any means within the resources of nature, can extract an unlimited amount of water from a limited amount of air within a limited time. In the month of June, up to and including the date last given, vast quantities of rain had fallen on ground occupied by the contending armies; and we might naturally expect that the atmosphere would, for a time, respond slowly to further calls upon it. Yet it required but a little time to receive new accessions of vapor for the production of other great rains, as will be seen.

The famous

SEVEN DAYS' FIGHT

commenced in the afternoon of the twenty-sixth of June, 1862, with the battle of Mechanicsville,[65] though there was an affair on the preceding day that involved a loss of some five hundred men in killed, wounded and missing. On the twenty-seventh was fought the sanguinary

BATTLE OF GAINES' MILL,

and on the twenty-eighth there was considerable artillery firing, but no regular battle. The twenty-seventh was a bright clear day, as was also the twenty-eighth, except that on the morning of the twenty-eighth there was, for a time, an appearance as of coming rain; but on the night of the twenty-eighth and morning of the twenty-ninth, it rained

---

[65] For the authority for this division, see Greeley's American Conflict, Vol. II., page 167.

heavily.[66] This rain appears to have been confined to a comparatively limited extent of country.

On the twenty-ninth was fought

THE BATTLE OF SAVAGE'S STATION.

A heavy thunder storm followed in the night, passing over a part of the country East of and near the battle-field, though perhaps not reaching the field itself. At a point on the Pamunkey River, between White House and the York River, the storm lasted with varying severity for from five to six hours, during two of which the rain fell in torrents.[67]

Next day was fought the battle of Glendale, and on the next day after, (July 1, 1862), the fearful

BATTLE OF MALVERN HILL.

A terrific storm followed, commencing before daylight the next morning,[68] and continuing through the day, and during most of the following night; and accompanied, during a portion of its progress, with hail,[69] as well as with thunder and lightning, and torrents of rain. This storm appears to have extended over all the surrounding country.[70]

This day of storm (July 2, 1862), was the last of the historic seven of battle and retreat; but after the battle of Malvern Hill there is no account of

---

[66] See Document No. 26.

[67] See Log of U. S. Steamer Sebago on Pamunkey River.

[68] See Documents Nos. 1, 20, 26 and 41.

[69] Log of Steamer Sebago in Hampton Roads.

[70] See Log of Steamer Sebago in Hampton Roads, and of the "Galena" on James River.

further canonading until the morning of the 3rd, when, at half-past 10 A.M., the rebels commenced

THROWING SHELLS INTO THE CAMP,

at Harrison's Bar, but were soon driven off by the fire of the batteries and of the gunboats. In the evening and night of that day it again rained.[71]

While the army remained inactive at Harrison's Bar, after the above battles, there occurred an instance of

GUNBOAT FIRING ON JAMES RIVER,

followed by rain. In the journal to which reference has been made, is written, under date of July 15, 1862; "there has been some gunboat firing down the river." Also, under same date, "at dark a heavy thunder storm."[72]

Having shown that all the great battles of Gen. McClellan's campaign against Richmond were followed by great rains, and most of the minor collisions by rains more than proportionately heavy, I will next show that rain also followed all the principal engagements of the army of Virginia, commanded by Maj. Gen. Pope, which soon after the date last above referred to, advanced against the enemy in Virginia, by way of Culpepper Court House, on the Orange and Alexandria Railroad. The first engagement of this campaign was

[71] Log of the U. S. Steamer Galena on James River.
[72] See Doc. No. 20.

THE BATTLE OF CEDAR MOUNTAIN,

fought August 9, 1862, between Gen. Banks' corps and a superior force under Stonewall Jackson. Rain followed as usual;[73] but as the amount of artillery firing in this engagement was small, so likewise was the amount of rain which it apparently produced, a certain proportion being observed between the two as compared with some other battles.[74]

The next engagement of any consequence was at the Rappahannock River, on Gen. Pope's retreat, and consisted principally in

HEAVY ARTILLERY FIRING AT KELLEY'S FORD AND RAPPAHANNOCK STATION,

on the 20th, 21st, and 22nd August, it being particularly heavy on the 21st and 22nd. In the night of the 22nd a tremendous rain set in, which drowned all the fords, and carried away all the bridges at the front, and rendered impossible an aggressive movement which Gen. Pope had meditated.[75] There was also a shower in the afternoon.

The next heavy artillery firing was on the night of the 26th, followed by still more on the 27th. During this day, different portions of Gen. Pope's forces were engaged with the enemy, the most serious encounter being the

[73] See Doc. Nos. 11 and 38.
[74] See Doc. No. 38.
[75] Am. Conflict, Vol. II, p. 178.

FIGHT AT BRISTOW STATION,

in which there was a loss of some three hundred men on each side. This was followed, at about 9 o'clock in the evening, by a little rain, and on the day following by a heavy shower.[76]

We come next to the

SECOND BATTLE OF BULL RUN,

fought on the 29th and 30th of August, 1862. The battle commenced in the morning of the 29th, and was followed on the morning of the 31st and afternoon of the next day by heavy rains.[77]

The last of this series of engagements was the

BATTLE OF CHANTILLY,

fought September 1, 1862. It was commenced at 5 P. M., by two divisions under Gen. Reno, which attacked a superior force under Stonewall Jackson, and were repulsed. Afterwards, Gen. Kearney " advanced and renewed the action in the midst of " a thunder-storm so violent that ammunition could " with great difficulty be kept serviceable, while " the roar of cannon was utterly unheard at Centre- " ville, barely three miles distant."[78] To the cannonading on the last day of the preceding battle this storm should, perhaps, in a great measure be attributed.

---

[76] See Docs. Nos. 20 and 38.
[77] See Docs. Nos. 1, 5, 11, 20, 22, and 38.
[78] Am. Conflict, Vol. II. p. 188. See also Docs. Nos. 11 and 20.

At the

GREAT BATTLE OF ANTIETAM,

in Maryland, the phenomenon of rain following the discharge of artillery was again exhibited. The battle was fought on the 17th of September, 1862; the rain was on the afternoon of the 18th, and consisted of a sudden and heavy shower.[79]

THE BATTLE OF PERRYVILLE, OR CHAPLIN'S CREEK, KENTUCKY,

fought between the armies of Generals Buell and Bragg, on the 8th of October, 1862, furnishes a remarkable instance of rain following artillery firing, during a time in which the state of the atmosphere would be considered by some as exceedingly unfavorable to the production of that phenomenon. A great drouth was prevailing in the State at that time, causing severe privation and suffering in the army both to men and animals;[80] but the battle seems to have brought a change, for a heavy rain followed.[81] This fact is important, as it shows that a state of drouth by no means proves that there are not ample supplies of aqueous vapor somewhere within reach of the noise and concussion produced by the discharge of ordnance, and which can be drawn on for rain at any time.

THE BATTLE OF PRAIRIE GROVE, ARK.,

fought Dec. 7, 1862, furnishes a somewhat similar

---

[79] See Appended Documents, Nos. 1, 2, 22, and 33.
[80] Am. Conflict. Vol. II. page 218.   [81] See Doc. No. 8.

instance. We read in history that the weather at the time was clear and dry;[82] and yet we learn that on the day after the battle it rained.[83]

The firing at the

CAPTURE OF VAN BUREN, ARK.,

was also followed by rain.[84]

A heavy storm followed the

ATTACK ON THE DEFENCES ON THE NORTH SIDE OF VICKSBURG,

by the formidable expedition that was sent against that place in December, 1862, under General Sherman and Admiral Porter. From the commencement of the debarkation of the troops, on the morning of the 26th, until the battle, the weather was good, being for the most part "clear and pleasant." There was some preliminary fighting on the 28th, and on the 29th the grand assault was made, the battle commencing early in the day. Between four and six in the evening rain commenced to fall, and from eight to midnight it came down in torrents. This rain continued until about eight o'clock the next morning.[85]

THE BATTLE OF MURFREESBORO, OR STONE RIVER,

is one of the many great battles that have commenced in fine weather and ended in pouring rain.

[82] American Conflict, Vol. II. page 37.
[83] See Docs. Nos. 4 and 34.
[84] See Doc. No. 4.
[85] Log of the Benton. See also, Am. Conflict, Vol. II. p. 291.

This battle was fought on the 31st of December, 1862, and 1st and 2nd of January, 1863. The first day of the battle was bright[86] and clear, but on the last a heavy storm set in, which continued through the night and a great part of the following day.[87]

We have seen that nearly all[88] the battles, both great and small, of the Eastern armies, up to and including that of Antietam, were followed by rain. After Antietam, the next great battle fought by the Army of the Potomac was the

BATTLE OF FREDERICKSBURG, VA.,

fought Dec. 13, 1862, and the same is true of this as of the others. The day of the battle, with the exception that there was a fog in the morning, was bright and sunny,[89] but a heavy storm of rain followed, commencing on the night of the 15th, while the army was re-crossing the Rappahannock.[90]

Next in order of the battles of the Army of the Potomac was the

BATTLE OF CHANCELLORSVILLE,

fought May 2nd, 3rd and 4th, 1863, and at this, too, the same phenomenon was exhibited. On the third day after the commencement of the movement, in the midst of a rapid cannonade, there came on a fear-

---

[86] Greeley's American Conflict, Vol. II. p. 279.

[87] See Docs. 8, 14, 16, 28 and 29. Also, Am. Conflict, Vol. II, p. 280.

[88] The battle of South Mountain, Md., fought Sept. 14, 1862, seems to have been the most marked exception to the general rule.

[89] Am. Conflict, Vol. II, p. 344.

[90] See Docs. 1, 2, 13, 22, 31 and 38.

ful thunder storm, and for a time the soldiers fighting in the woods were unable to distinguish the "artillery of heaven" from that of earth.[91] In the afternoon and night of May 5, the storm was so violent as to cause a great flood in the Rappahannock, sweeping away some of the pontoons forming the bridges on which the army was that night re-crossing the river; thus delaying the movement and threatening for a time to lead to serious consequences.[92]

It also rained[93] immediately after the

BATTLE OF BEVERLEY FORD, VIRGINIA,

fought June 9, 1863. This was a sharp fight, lasting about half a day, the forces engaged on the Union side consisting, besides cavalry, of two brigades of infantry and two batteries of artillery detached from the Army of the Potomac.

Following Chancellorsville, the next great encounter of the Army of the Potomac with that of General Lee was

THE BATTLE OF GETTYSBURG, PA.,

fought July 1, 2 and 3, 1863; and this, too, was followed by a rain, and one that one would compare, in the amount of water that fell, with the rains which had followed any of the previous battles. The battle was fought in clear weather, except that during the first day's fight there was a slight shower,

[91] See Doc. No 24.
[92] See Docs. Nos. 1, 2, 12, 22, 25 and 38.
[93] See Doc. No. 11.

and again another in the evening of that day, but they were both so unimportant as to have been generally unnoticed. The great rain commenced on the night of the 3rd, about six hours after the firing had ceased; and through the 4th, and also part of the 5th, it rained furiously. The storm must also have extended a great distance southwestward, as it caused a flood in the Potomac which lasted several days, stopping, in the meantime, the retreat of the rebel army.[94] At Westminster, about thirty miles in a southeasterly direction from the battle field, the rain seems to have commenced about eighteen hours later than at the latter place; and it continued to rain there heavily through the second night after battle.[95]

After the return of the Confederate army to Virginia, pursued by the Army of the Potmac, rain still continued to follow their battles. The

ENGAGEMENT NEAR BRISTOW STATION

may be mentioned as an instance. A former engagement at that place has already been referred to. The second was a fight which occurred on the 14th of October, 1863, between portions of the respective armies, in which six pieces of artillery were captured from the Confederates, while the loss of the Union side in killed and wounded was about 200 men. This, on the 16th, was followed by a heavy

[94] See Docs. Nos. 2, 3, 5, 11, 20, 24, 26, 31 and 38.
[95] See Doc. No. 12.

rain, rendering the creeks unfordable, and seriously interfering with the plans of the Union commander.[96] The affair of

MINE RUN, VA.,

is another instance. This movement took place in November, 1863. The heaviest fighting was on the 27th, being such as to entail a loss on either side of from 300 to 500 men in killed and wounded. The next day at evening there was a pelting rain.[97] The

DESTRUCTION OF THE REBEL STEAMER, NASHVILLE,

near Fort McAllister, Ga., by the U. S. vessels, Montauk, Seneca, Wissahicon, and Dawn, furnishes a good instance of heavy rain apparently brought on by an action in which only a moderate number of guns were employed. On one side were the Union vessels named, which fired deliberately, and on the other the Nashville and the fort. The engagement took place on the 28th of February, 1863, and lasted two hours and three-quarters. The following is from the log of the steamer Montauk:

February 28, from 12 to 4, A.M., "Light, variable airs and clear weather." * * * "At 7.07 opened fire on the Nashville, aground in 7 mile reach."

From 12 to 4, P.M., "Light easterly winds and partially overcast."

---
[96] Am. Conflict, Vol. II. p. 396.
[97] American Conflict, Vol. II. p. 401.

From 6 to 8, P.M., "Moderate wind from S.W.; cloudy and rainy."

From 8 to 12, midnight, "Light baffling winds and much rain."

February 29, from 12 midnight to 2 A.M., "Incessant rain."

March 1, 8 to 12 A.M., "Pleasant."

THE BATTLE OF CARNEY'S BRIDGE, LA.,

fought January 14, 1863, and in which four gunboats and four or five regiments of troops were engaged, was followed in the night by a furious rain, which commenced about 1 A.M., and continued, with varying severity, until 8. There had been some rain the night before, and the morning of the action was cloudy.[98]

At

PORT HUDSON,

on the Mississippi River, a number of naval and military engagements occurred that were each followed by rain. The first that will be mentioned was the

PASSAGE OF THE BATTERIES

by Admiral Farragut, with a number of vessels of his fleet, on the night of the 14th of March, 1863. Fire was opened at about half-past 11, and " soon " the earth trembled to the roar of all the rebel " batteries."* A vast bonfire was kindled, by the

---

[98] Log of the "Calhoun."

* Am. Conflict. Vol. II. p. 329.

light of which the rebel gunners poured their fire into the passing vessels, while the latter replied with broadside after broadside, as each came within range. This commotion of earth and air was not without its effects. The weather of the day preceding and morning following the action showed blue sky, with detached clouds; and at the commencement of the fight there was a light breeze blowing from the northward. Soon after the battle commenced, however, it became calm, and so continued until about 9 o'clock the next evening. But before this time the storm had commenced—coming up between 12 and 1 in the afternoon. At 1 it rained, and at 2 it poured. From this time until 10 o'clock at night it rained incessantly, the rain, until 8, falling in torrents.[99]

The

ASSAULT ON PORT HUDSON,

by Gen. Banks, May 27, 1863, was also followed by heavy rain. The sky, on the morning of that day, was cloudless, but on the 29th it rained heavily and continuously for four hours.[100]

Again, on the 9th of June, 1863, during a spell of clear and pleasant weather, it is recorded in the log book of the "Hartford," that heavy firing was heard at Port Hudson. The next morning the sky became overcast, but the clouds afterwards dispersed, and from 4 to 8 A.M. it was again " clear and

---
[99] Log of the U. S. Steam Sloop "Hartford."
[100] Log of the Hartford. See also Doc. No. 41.

pleasant." But, on the morning of this day (the 10th) an attempt was made by Gen. Banks, under a

HEAVY FIRE OF ARTILLERY,

to establish his lines within attacking distance of the enemy's works.[1] The firing was heavier than that of the day before, and within less than twenty-four hours it was followed by floods of rain. The log of the Hartford for the 11th says: "About 3.20 A.M., squall of wind; let go the port anchor; rain came up from northward, and continued to blow 15 minutes, and rain until 4 A.M." (the end of the watch). The officer of the next watch (from 4 to 8 A.M.) enters in the log the following: " Heavy firing at Port Hudson during the watch, also heavy rain."

This latter

FIRING AT PORT HUDSON

was followed, after a cessation of the above storm for some hours during the middle of the day, by a violent shower in the latter part of the afternoon.[2]

Again, at the

SECOND GENERAL ASSAULT UPON PORT HUDSON

the same phenomenon was repeated. This assault was delivered on the 14th of June, 1863. The weather on the 12th had become " clear and pleasant, with light breeze from northward; " but

---

[1] Am. Conflict. Vol. II. p. 335.
[2] Log of the Hartford.

on the 16th it again rained heavily, with thunder and lightning, and with squalls of wind as before.³

I have mentioned the bombardment and passage of the Vicksburg batteries on the Mississippi by Admiral Farragut, on the night of June 28, 1862, as an engagement followed soon by a storm of rain with heavy thunder and lightning; and his passage of the Port Hudson batteries has also been referred to as succeeded by a tremendous and long-continued shower. It remained for Admiral Porter to try, at the former place, a similar experiment. We might naturally expect that, if Farragut could bring rain by steaming past rebel batteries and engaging them as he passed, Admiral Porter could do the same, and so it proved. As an exploit of war, the passage of the Vicksburg batteries by Porter equalled that of Farragut, performed at an earlier period. As a scientific experiment for the artificial production of rain, it was still more successful.

PORTER'S PASSAGE OF THE VICKSBURG BATTERIES

was made on the night of April 16-17, 1863. Eight gunboats passed down, and when opposite to the city, "in a moment the whole bluff was ablaze with the flashes and quaking to the roar of heavy guns rising, tier above tier, along the entire water front of the city."⁴ The action lasted a little less than two hours, terminating at about 1 o'clock in the morning of the 17th. It occurred during a spell of "clear and pleasant" weather; but on the

³ Log of the Hartford. ⁴ Am. Conflict. Vol. II. p. 301.

18th, from 6 to 8 P.M., there was "rain at intervals," and from 8 to 12, "heavy squalls, with continuous thunder and lightning, and deluges of rain."[5] The rain continued to fall heavily until about 4 o'clock next morning.

The

BATTLE OF RAYMOND, MISS.,

fought May 12, 1863, was followed, on the 14th, between the hours of 9 and 11 A.M., by a tremendous shower.[6]

During the

SIEGE OF VICKSBURG, MISS.,

which commenced May 19, and ended July 4, 1863, there were numerous showers, though at that point they were not generally heavy. The following are some of the days on which it rained,[7] viz.:

May 22, 27, 28, and 31, June 10, 15, 16, 23, and 24. Besides the days on which there was rain, it was "cloudy, with appearance of rain," on the 21st of May; and cloudy at different hours on the following days, viz.: May 23 and 25, June 3, 5, 11, 17, 18, 19, 20, 21, 25, and 26; with "passing clouds" on the 27th and 30th. On these days it would generally become cloudy for only a few hours, and then clear off again.

While the circumstances attending this siege are

---

[5] Log of the U. S. Ram "Lafayette."

[6] Am. Conflict. Vol. II. p. 306.

[7] These dates and facts are taken from the Log of the "Blackhawk," except the first, which is from that of the "Benton."

not such as to afford strong evidence in support of the proposition, that artillery firing can at all times be made to bring heavy rain, neither do they furnish evidence to the contrary.

Rain followed the

NAVAL ACTION OFF CHARLESTON HARBOR,

that ensued on the morning of January 31, 1863, when two iron-clad Confederate vessels came out and attacked the Union blockading fleet. The weather previously had been clear, and so remained until 8 A.M. of the next day. It then began to get cloudy; at noon the sky was completely overcast; at 7.30 in the evening there was a "light sprinkling of rain," and from midnight to 8 A.M. the weather is described simply as "rainy."[8]

During other operations by the army and navy in front of Charleston, in the year 1863, engagement after engagement was followed by rain. Of a long series of fights, there were but two where the phenomenon was not exhibited, and these were followed by overclouded skies.[9]

THE ATTACK ON THE DEFENCES OF SECESSIONVILLE,

on James Island, by General Hunter, was one of those where storm quickly succeeded battle. The engagement took place on the morning of June 16, 1863, and eight hours of continuous rain followed it,

---

[8] Log of the "Keystone State."
[9] Attacks of April 7 and Sept. 5, 1863.

commencing between 7 and 8 o'clock in the morning of the 17th.[10]

The

ATTACK ON MORRIS ISLAND,

July 10 and 11, 1863, is another of the list. The bombardment, assault and capture of the batteries on the south end of the island was made on the 10th, and the unsuccessful attack on Fort Wagner on the 11th. The sky during the first day was cloudless; on the second it was cloudy in the morning and thickly overcast in the evening, and on the following night it rained with extraordinary violence.[11]

Another of this series of engagements was the

ATTACK ON GENERAL TERRY ON JAMES ISLAND,

made at daybreak on the morning of July 16, 1863, and which was repulsed by the aid of five gunboats, which happened to be near. This was only a few days after the storm just mentioned, but it had passed off and none but detached clouds were visible in the sky. It commenced clouding up, however, about 5 o'clock in the afternoon; between 8 and 9 in the evening it had become rainy and squally, and at 11 it commenced to pour in torrents. This storm, which continued the greater part of the night,[12] and, after an intermission through part of the night following, is spoken of in history as "terrible."[13]

[10] Log of the U. S. Steamer "Pembina."
[11] Log of the "Catskill."
[12] Log of the "New Ironsides."
[13] Am. Conflict, Vol. II. p. 476.

Next in the list is the

BOMBARDMENT OF FORT WAGNER,

on the 18th of July, 1863. The gunboats commenced firing at 8.30 in the morning, and the larger vessels and land batteries at 12.30. " On our side fully a hundred great guns steadily thundered. * * * As the day declined the roar of our great guns, no longer incessant, was renewed at longer and longer intervals, and finally ceased; our iron-clads, save the Montauk, returning to their anchorage; while a thunderstorm burst over land and sea, sharp flashes of lightning intermitting and intensifying the fast-coming darkness.."[14] * * * This storm continued until 4 o'clock the next morning[15]

Again, on the 20th of July, as shown by the log book of the " New Ironsides," there was

HEAVY FIRING ON FORT WAGNER.

Rain followed in the night of the 21st, also on the night of the 22nd and 23rd.

Again, on the 24th of July there was

ANOTHER ATTACK ON SUMTER, WAGNER AND CUMMINGS POINT BATTERIES.[16]

Heavy rain followed, commencing at 1 A.M. on the 25th, and continuing until 11 A.M.[17]

---

[14] Greeley's American Conflict, Vol. II. p. 476.
[15] Log of the "New Ironsides."
[16, 17, 18 and 19] Log of the "New Ironsides."

Again, on the 28th of July there was

MORE HEAVY FIRING,

the "James Island batteries firing on our batteries, our mortar batteries firing on Fort Wagner."[18] Rain followed the next day at 2 A.M.[19]

The next engagement of this series was the one that ensued

WHEN GEN. GILMORE'S SIEGE BATTERIES OPENED FIRE

on Sumter, Wagner, and the Cummings' Point batteries. This was on the 17th of August, 1863, commencing at a very early hour in the morning. There was a light wind at the time from the northwest, and the sky showed blue with detached clouds. At 7 A.M. the wind became variable, and at 2 P.M. it blew lightly from the southeast; at 6 P.M. the rain began to fall, and for four hours it poured without intermission.[20] The wind changed at 7 P.M. to the northeast, but it blew gently all that day, though we read in history that on the 18th and 19th a heavy northeaster raged.[21]

I do not doubt that my readers are wearied with the sameness of this recital, but I am not yet done even with the list of engagements before Charleston On the 23rd of August, commencing at 3.15 and lasting until 6.30 in the morning, there was an

[20] Log of the "New Ironsides."
[21] Am. Conflict, Vol. II. p. 479.

ATTACK BY FIVE MONITORS ON FORT SUMTER, MOULTRIE REPLYING.

There had been no rain since the storm last chronicled, nor was there apparently any indications of rain when the action commenced, though there was that morning a fog; but within less than twenty-four hours the sky became overclouded, and in another hour it rained.[22]

The history of

ROSECRANS' ADVANCE

from Murfreesboro, Tennessee, furnishes a further instance of the remarkable connection between military operations and rain. There was a good deal of artillery firing in this movement,[23] and for seventeen successive days it rained every day.[24] The engagement at

LIBERTY GAP,

fought about the 24th of June, 1863, was followed by heavy rain.[25] The same is true of the[26]

BATTLE OF SHELBYVILLE,

fought on the 27th June, 1863. After this engagement, Elk River became so swollen as to stop for some days the pursuit of the retreating Confederates.[27]

[22] Log of the "New Ironsides."
[23] See Doc. No. 16.
[24] Am. Conflict, Vol. II. p. 409.
[25] See Doc. No. 29.
[26] See Doc. No. 14.
[27] See Am. Conflict Vol. II. p. 410.

The following may or may not be an instance worth recording of the occurrence of rain following the discharge of artillery. At Tebb's Bend, on Green River, in Kentucky, on the 4th of July, 1863, the rebel General Morgan, with a force of two regiments and four guns, made a desperate but unsuccessful attack on a Union force under Col. O. H. Moore, which lasted for several hours. The next day he spent seven hours, commencing at sunrise, in endeavoring to

REDUCE THE DEFENCES AT LEBANON.

A rain followed, for we read that he finally charged into the place, set it on fire and compelled its surrender; and that at dark a furious rain came on, during which he raced his prisoners ten miles in ninety minutes to Springfield—all except one, who being unable or unwilling to keep up with the rest, was shot.[28]

A rain occurred also after

THE BATTLE OF CHICKAMAUGA, GA.,[30]

fought Sept. 19 and 20, 1863. A circumstance connected with the weather noticed after each day of battle was a dense fog; the one on the morning of the 20th was so thick that objects could scarcely be distinguished at a few steps distance.[31] The battle was fought in the woods, where but little

---

[28] Am. Conflict, Vol. II. p. 405.
[30] See Doc. No. 14.
[31] See Am. Conflict, Vol. II. p. 419. Also Doc. No. 8.

PRODUCTION OF RAIN. 51

artillery could be used, and where we might expect that the effect of concussion would be lessened by the interference of the trees with the movement of the air. The precise time when the rain occurred is not stated, but it is probable, from the reasons above given, that it was a little longer in "brewing" and less in quantity than the rains which generally follow great battles—a supposition to which additional probability is given by the fact that, by some who were present, this rain is not remembered.

At the

BATTLE OF LOOKOUT MOUNTAIN,

fought on the 23rd and 24th of November, 1863, a circumstance occurred of a similar nature, and fully as remarkable as would have been the production of rain. On the 24th "darkness at 2 P.M. arrested our victorious arms, the mountain being now enveloped in a cloud so thick and black as to render further movement perilous, if not impossible."[32]

THE BATTLE OF MISSION RIDGE,

fought the next day, was followed by rain.[33]

During Gen. Banks'

RED RIVER CAMPAIGN,

in the spring of 1864, in which there was more or less fighting daily for several weeks, there was much showery weather,[34] but the precise dates on which

---

[32] Am. Conflict, Vol. II. p. 439.
[33] See Doc. No. 29.
[34] and 36 See Doc. No. 41.

rains occurred I have not been able generally to ascertain. Probably many of them were showers which extended over only a limited space of country —as on the 7th of April we read that a heavy rain occurred which greatly retarded the rear of his extended column but did not reach its front.[35] A

FIGHT ON THE ATCHAFALAYA RIVER

has been mentioned[36] as one where the phenomenon was specially noticeable, from the weather previous to the fight having been so clear and bright. The

ENGAGEMENT NEAR MARKSVILLE

(or Mansura), which took place May 16, 1864, was followed by nearly a week of rain.[37]

GEN. STEELE'S CAMPAIGN IN ARKANSAS,

made while Gen. Banks was operating in the adjoining State, was also attended with heavy rains,[38] some of which, it is not impossible, may have had their origin in the part of the country then occupied by the latter. The

BATTLE OF MARKS MILL,

fought by a portion of his command on the 25th of April, 1864, was followed by rain, for we read that "by daylight of the 27th his army was across the Washita and in full retreat amid constant rains."[39]

[35] Am. Conflict, Vol. II. p. 539.
[37] See Doc. No. 41.
[38] Am. Conflict, Vol. II. p. 552.
[39] Am. Conflict, Vol. II. p. 553.

On

SHERMAN'S ATLANTA CAMPAIGN,

which was a continuous battle for ninety days, there were heavy rains at short intervals.[40]

AT THE BATTLE OF DALLAS, GA.,

fought May 26, 1864, the circumstance was specially noticeable.

Great rains followed most of the battles of Gen. Grant's campaign against Richmond. The first engagement which took place upon his advance across the Rapidan was the

BATTLE OF THE WILDERNESS,

fought May 5 to 9, 1864, the heaviest fighting being on the 5th and 6th, and being for the most part an infantry battle, as it took place in the woods, where artillery could not be used to advantage. A little rain appears to have fallen[41] on the 8th or 9th, which increased to heavy thunder-storms after the first day of the

BATTLE OF SPOTTSYLVANIA COURT HOUSE,

which was a continuation of the Wilderness battle, and one in which much artillery was brought into action. This terrible battle was fought on the 10th, 11th, and 12th. Heavy rain set in on the night of the 10th.[42] On the afternoon of the 11th it also

[40] See Docs. Nos. 3, 8 and 29.
[41] See Doc. No. 26.
[42] See Doc. No. 33.

rained heavily. On the morning of the 12th there was a fog of exceeding density, and at noon rain set in again and fell in torrents, accompanied with thunder and lightning.[43] This storm extended over a hundred miles southeastward, and there lasted, with varying intensity, until midnight of the 13th.[44]

Gen. Butler's

BATTLE OF BERMUDA HUNDREDS,

fought on the morning of May 16, 1864, was followed by rain in the evening.[45] There was also rain on the 18th and 19th, on both of which days there was fighting along his front.

Gen. Grant's

BATTLE OF NORTH ANNA RIVER,

which was the next battle of his campaign after that of Spottsylvania, was followed by a heavy storm of rain, accompanied with thunder and lightning.[46] The battle was fought on the 23rd of May, 1864, and the storm commenced the day after,[47] and lasted during portions of three days.[48]

A spirited

FIGHT AT HAWES SHOP,

which occurred on the 28th of May, and in which

---

[43] See Docs. Nos. 2, 25, 33. Also Am. Conflict, Vol. II. p. 571.

[44] Log of the U. S. Steamer, "Commodore Perry," on James River.

[45] Log of the U. S. Steamer, "Agawam," on James River.

[46] See Doc. No. 25.

[47] Am. Conflict, Vol. II. p. 579. Also Log of Steamer Agawam.

[48] Log of U. S. Steamer Agawam.

the aggregate loss on both sides was some 1,200 men, was followed in the night of that day by rain[49] on James River.

Tremendous rains accompanied and followed the

BATTLE OF COLD HARBOR OR BETHESDA CHURCH,

which was the next engagement of this campaign. This fearful battle was fought on the 1st, 2nd, and 3rd of June, 1864; the commencement, on the 1st, being at 4 o'clock in the afternoon. So far as can be known from the state of the weather some fifty miles southeastward, there had been no rain since the night of May 28; and the battle was commenced under a cloudless sky.[50] But on the night of the 2nd there was a heavy rain; the next night there was another, and the third day of battle was followed by a third. Each separate day's encounter seems to have been followed by a separate rain, and the last—the one in which, for a time, the fighting was so furious, that in the space of twenty minutes "fully ten thousand of our men were stretched writhing on the sod, or still and calm in death"[51]—was followed by one of some twenty-four hours' duration, commencing in the afternoon of the succeeding day.[52]

The following are some of the other engagements of this campaign that were followed by rain, viz.:

[49] Log of U. S. Steamer Agawam.
[50] Log of the "Agawam."
[51] Quoted from Am. Conflict, Vol. II. p. 582.
[52] Log of the U. S. Steamer "Agawam," on James River. See also Doc. No. 2.

FIGHT AT BAILEY'S CREEK,

August 12, 1864. Rain followed on the 14th.[53]

A SECOND ENGAGEMENT AT BAILEY'S CREEK,

August 16, 1864. Rain followed on the 17th.[54]
Battle for the possession of the

WELDON RAILROAD,

fought August 18, 1864. Thunderstorm followed in the night.[55]

Two other

ENGAGEMENTS ON THE WELDON RAILROAD,

August 21, 1864. Same remark as to the above.[56]

ASSAULT AND CAPTURE OF FORT HARRISON,

September 29, by General Butler. Rain next afternoon[57] on James River.

ATTEMPT TO RETAKE FORT HARRISON,

by the rebels, September 30, 1864. Rain in the night and next forenoon on James River.[58]
Battle on the

SQUIRREL LEVEL ROAD.

Heavy rain immediately after.[59] At a point on the James River, the shower occurred between 8 P.M. and midnight of the same day.[60]

[53, 54] Log of the Steamer Agawam.
[55, 56] See Docs. 25 and 36.
[57, 58, 60, 61, 63] Log of the Steamer Agawam.
[59] See Doc. No. 24.

ACTION AT THE FRONT,

October 2, 1864. Rain next afternoon.[61]

BATTLE OF HATCHER'S RUN,

fought October 27, and the last one of Grant's battles for the year 1864. A heavy storm followed, accompanied with thunder and lightning.[62] At a point on the James River the rain poured for seven hours[63] during the night after the battle.

In West Virginia, the

FIGHT AT DUBLIN BRIDGE,

May 10, 1864, was followed by a fall of rain.[64]
In the Virginia Valley, the

BATTLE OF NEWMARKET,

fought May 15, 1864, was also followed by rain.[65]
There was some

CANNONADING AT MARYLAND HEIGHTS,

on the night of July 6, 1864, and sharp fighting on the 7th, was followed, on the night of the 7th, by a little rain, and on the next night by an "awful rain." The previous weather had been very dry.[66]

The

BATTLE OF WINCHESTER

(Crook's), fought July 24, 1864, was also preceded

---

[62] See Docs. Nos. 24 and 25.
[64] See Doc. No. 17.
[65] See Doc. No. 23.
[66] See Doc. No. 18.

by a long spell of dry weather; but the next day there was a hard rain.[67]

The general character of the weather in the Shenandoah Valley in the months of August and September, 1864, was that of drouth, and it is only remembered as such by an officer who has favored me with a communication upon the subject.[68] Yet an actual record that was kept of the weather in that section, during a part of the time mentioned, shows frequent instances of

RAIN FOLLOWING ARTILLERY SKIRMISHING.[69]

I quote:

August 17. " Clear at daylight." " Heavy fire of artillery " during the day.

August 18. " Rain."

August 19. " Skirmishing near Berryville."

August 20. " Rain."

September 3. " Cloudy; heavy artillery and musketry in the direction of Berryville." " Rain."

On the same day. " Still fighting far away into the night." Next day, " Rain."

Skirmishing also on the 4th. Rain on the 5th.

Also on the 5th, " Skirmishing heavy." September 6, " Rain all day."

September 9, " Smart skirmishing." September 10, " Rain." September 12, " Rain."

September 13, " Clear; " " Cannonading heavy." September 14, " Rain; " 15, " Cloudy; " 16, " Rain."

[67 and 68] See Docs. Nos. 17, 18, and 23.
[69] See Doc. No. 18.

The above memoranda were made by a Confederate officer, who was killed at the battle of Opequan Creek, near Winchester, Sept. 19, 1864, and some extracts from whose diary have been published.[70]

What entry he would have made had he lived after that battle, and the battle of Fisher's Hill, fought three days later, remains a mystery; but in the southeastern part of the same State (Virginia) it rained hard soon after each of those battles — the first rain occurring on the night of the 21st and the second about the middle of the day on the 23d.[71]

At the naval action and the bombardment of the forts at the entrance of Mobile Bay, further instances of the phenomenon under consideration were exhibited. There had been a shower on the day preceding the commencement of operations, but the weather on the evening of that day, and at the commencement of the first engagement, was such as the words, " blue sky with detached clouds," are used in the navy to describe. The

PASSAGE OF THE FORTS AND BATTLE WITH THE GUNBOATS,

which took place on the morning of Aug. 5th, commencing at 6.45 A.M., was followed by about two hours of rain, which commenced at 9 o'clock A.M.[72]

[70] In Putnam's "Record of the Rebellion," Vol. XI. p. 153.
[71] Log of the Steamer "Agawam " on James River.
[72] Log of the U. S. Steam Sloop " Hartford."

The
BOMBARDMENT OF FORT GAINES,
on the 6th, was followed by a thunder-shower on the 8th.[73]

The
BOMBARDMENT OF FORT MORGAN,
on the 9th, was followed by two excessively stormy days on the 10th and 11th, on which there was rain, thunder and lightning and squalls of wind.[74] The rain which fell at this time is described by an officer who was present as exceedingly copious.[75]

THE BATTLE OF FRANKLIN, TENN.,
was followed by a rain that froze as it fell, and covered the country with ice.[76] This battle was fought Nov. 30, 1864.

The
BATTLE OF THE CEDARS
was also immediately followed by rain.[77] This battle was fought shortly after the battle of Franklin, by troops under Gen. Milroy, sent from Murfreesboro to the relief of Fortress Rosecrans, invested by the rebels Dec. 4.

THE GREAT BATTLE OF NASHVILLE,
fought Dec. 15 and 16, 1864, was followed by one

[73 and 74] Log of the U. S. Steam Sloop "Hartford."
[75] See Doc, No. 4.
[76 and 77] See Doc. No. 5.

of the most tremendous rains that have ever been noticed in connection with military operations.[78] For several days the rain fell incessantly. "The country was flooded; the brooks were raging rivers," and "the roads were hardly passable in the rear of the fleeing foe."[79]

At the

BOMBARDMENT OF FORT FISHER,

Dec. 24 and 25, 1864, we again find rain following heavy explosions. The experiment which was tried of exploding a ship-load of powder under the walls of the fort, took place at 1.45 A.M. of the 24th, and the bombardment by the fleet commenced at half-past 11 A.M., and continued during the remainder of the day. The next day it was renewed for seven hours. On the first day there were no indications of rain, nor had there been on the day preceding; but on the second day, at 1 o'clock in the afternoon, it became cloudy, and at 7 in the evening rain commenced, which continued through the greater part of the night. During the first part of the night it only drizzled, but in the latter part for two hours it rained heavily. The morning that followed was rainy and squally, and after an intermission rain fell again in the middle of the day for about an hour.[80]

---

[78] See Docs. 5, 28 and 29.
[79] Am. Conflict, Vol. II. p. 687.
[80] Log of the U. S. Steamer "Malvern."

The

SHELLING OF THE WOODS

by the gunboats on the 27th, during the re-embarkation of troops belonging to the expedition, whose operations at Fort Fisher have just been noticed, was followed by more rain on the 28th, though on the 27th there had been none.[61]

The expedition against Fort Fisher above referred to will be remembered as the one which was unsuccessful. But the

SECOND EXPEDITION AGAINST FORT FISHER,

and the one that effected its capture, met, in the character of the weather that attended its operations, a somewhat similar experience. In the latter case, however, the storm was principally hail instead of rain. The operations of the fleet commenced in a heavy bombardment in the night of January 12, 1865, to cover the landing of the troops. There were no clouds during the day, except detached ones, and from 9 P.M. until midnight the sky was perfectly clear. But in the afternoon of the 13th, there was a hail storm, commencing at 2 P.M., and lasting, with an intermission of two hours, until midnight, after which it broke away.[62]

THE FIRING IN THE NIGHT OF THE 13TH

was followed by a second change of weather on the

[61] Log of the Steamer "Malvern."
[62] Log of the U. S. Steamer, "Malvern."

14th, the sky becoming overcast for four hours;[83] but the weather afterwards returned to its normal condition—"blue sky with detached clouds."

The bombardment, assault, and capture of the fort on the 15th, was followed, on the 16th, by two hours of drizzling rain.[84]

THE BATTLE OF AVERYSBORO, N. C.,

fought March 16, 1865, was also followed by rain. History relates that at the close of this battle, "Night fell dark and stormy."[85]

The

BATTLE OF BENTONVILLE, N. C.,

fought March 18, 1865, was likewise followed by rain, which on the 21st was heavy.[86]

We have seen that at the bombardment and capture of the lower forts at Mobile Bay, rain followed each operation of the fleet. The same is true of the operations of the army of Gen. Canby, and fleet of Rear-Admiral Thatcher, in the

REDUCTION OF THE UPPER FORTS AND CAPTURE OF THE CITY OF MOBILE,

months of March and April, 1865.

The preliminary firing by the army and navy in approaching the city was followed by rain on the evening and night of March 27.[87] The nature of

---

[83, 84] Log of the U. S. Steamer, "Malvern."
[85] Am. Conflict, Vol. II. p. 707.
[86] Am. Conflict, Vol. II. page 708.
[87] Log of the U. S. Steamer "Octorara."

some of this firing is shown by the following extract from the New York *Times* of April 7, 1865, viz.: " Gen. Steele's command met with much opposition, but no regular battle was fought until at Mitchell's Fork, on the morning of the 27th, where the enemy, numbering 800, made a stand, and after a severe fight were repulsed." There was also some gunboat firing about this time, and two of our vessels were blown up by torpedoes and destroyed. Previous to this rain there had been none for five or six days at the place or places of observation of the steamer " Octorara," one of the vessels engaged in the operations.

THE SIEGE OF SPANISH FORT

was opened on the 28th. The firing of this day and the next was followed by a heavy thunderstorm in the night of the 29th, accompanied with squalls of wind.[88]

Several other showers occurred, the last being on the 9th, the last day of the siege, and following the tremendous fire which was concentrated on Spanish Fort at nightfall on the previous day, and which effected its reduction.

The rain which followed the various engagements of this and the former expedition against Mobile, is described by an officer who was present, as more copious than any he had ever before witnessed.[89]

[88] Log of the Octorara.
[89] See Doc. No. 4.

THE BATTLE OF DABNEY'S MILL, VA.,

fought February 6, 1865, was followed next morning by a fall of snow.*

At the renewal of active operations before Richmond in the spring of 1865, storm still followed battle.

THE GENERAL ADVANCE BY GRANT'S ARMY

was made on the 29th of March, 1865, on which day was fought the battle of Quaker Road, and all that night and next day rain fell heavily.[90]

The last instance of the kind drawn from the late war of the rebellion which I shall mention is the

BATTLE OF FIVE FORKS,

immediately preceding the capture of Richmond. This battle, fought March 31, 1865, was, like so many others before it, followed by rain.[91] Before the battle commenced, the storm which followed the previous battle had ceased.[92]

So far I have only given instances of rain following the discharge of artillery occurring in the United States and Mexico, and in wars in which the army and navy of the United States took part. But in other parts of the world the same phenomenon has been noticed. The first instance that will be mentioned is one that was observed in the

---

* See Doc. No. 33.
[90] Am. Conflict, Vol. II. p. 731. See also Docs. Nos. 2 and 33.
[91] See Doc. No 2.
[92] Am. Conflict, Vol. II. p. 731.

harbor of Rio de Janiero, by one of our naval officers, some 27 or 28 years ago, on the occasion of the arrival there of the Princess of Naples, now Empress of Brazil. She was accompanied by the Neapolitan and Brazilian squadrons, and upon her arrival, the fortifications and foreign squadrons began to fire. The firing continued for an hour or more, when the sky was suddenly obscured, and heavy showers followed. Previous to this, the weather had been clear and beautiful. The next day was calm and partly overcast; as soon as the firing of salutes was resumed, the breeze sprang up, and the rain began to fall.[93]

THE BATTLE OF DRESDEN,

fought on the 26th and 27th of August. On the 27th the battle was renewed, under torrents of rain, and amid a tempest of hail. (Scott's Napoleon, chap. 27, p. 190.)[94]

THE BATTLE OF LIGNY,

fought on the 16th of June. "After the battle the weather was dreadful, as the rain fell in torrents." (Scott's Nap. chap. 27, p. 323.)[95]

THE BATTLE OF WATERLOO,

as all know, was fought in a pouring rain, brought on, without doubt, by the battles of Ligny and Gemappe, which preceded it.

[93] See Doc. No. 37.
[94] [95] Espy's " Philosophy of Storms."

PRODUCTION OF RAIN.

THE BATTLE OF EYLAU,

fought on the 8th of February. The action commenced at daybreak. Two strong columns advanced for the purpose of turning the right and storming the centre of the Russians, but they were repulsed in great disorder. The Russian infantry stood like stone ramparts, and kept back the enemy with a heavy and well-sustained fire from their artillery. About midday a heavy storm of snow commenced falling, which added to the obscurity caused by the smoke from the burning village of Serpallen. (Scott's Nap.)[96]

Capper on Monsoons, page 171, says:

AT MADRAS, ON THE 4TH OF JUNE, 1776,

morning fair, noon cloudy, in the evening rain. N. B. More than two hundred pieces of cannon fired in salutes; *quære*, whether it occasioned the rain? This *quære* is particularly appropriate, as this is the dry season on the Coromandel coast, and it did not rain after this till the 30th of the month.[97]

During the

SIEGE OF VALENCIENNES

by the allied army, in the year 1793, it rained violently every day soon after the heavy cannonading commenced. The allies employed 200 heavy ordnance, and the besieged had above 100, and they were frequently all in action at one time.[98]

[96, 97, 98.] Espy's "Philosophy of Storms."

M. Arago says, "I shall here repeat two facts which occur to my own memory, in the hope that they will lead to analogous statements. On the 25th of August, 1806, being the day selected for the

"ATTACK OF THE ISLET AND FORTRESS OF
DANNHOLM, NEAR STRAUSLAND,

"General Fririon, that he might harrass and fatigue the Swedish garrison, ordered it to be cannonaded during the whole day. In spite of these powerful and continued discharges of artillery, a violent thunder-storm visited the spot at 9 o'clock in the evening. Again it happened, oddly enough, that

THE ENGLISH LINE OF BATTLE SHIP, THE DUKE,

of 60 guns, was struck with lightning, in the year 1793, whilst it was cannonading one of the batteries of the Martinico."[99]

During the late war between France and Prussia, the occurrence of storms of rain after battles was specially noticeable, particularly so in the months of August and September; and the accounts from the battle-fields contained many allusions to the subject. For instance, immediately after the

BATTLE OF SEDAN,

at which the French Emperor was taken prisoner, we read, in a telegram dated Donchery, September 3, "It is raining torrents."

[99] Espy's "Philosophy of Storms."

Again, at the bombardment of Strasburg, in a dispatch dated Strasburg, September 8, *via* London, 10th, we read, "There are daily thunder-storms, and the Rhine has risen, driving the inhabitants from their cellars."

Such accounts, bringing to mind occurrences of a similar nature which took place in our own war, and which strongly impressed the writer at the time with the idea of the practicability of obtaining rain at will by the use of gunpowder, led him to believe that the time had come when some experiments ought to be made in the matter, other than those which are incidental to battles and sieges, and determined him to ask the co-operation of those who had observed the phenomena in question, in bringing the subject forward. The day after the first publication of an article written with this view, a letter appeared in the New York *Evening Post*, showing that the matter had already received much attention on the other side of the Atlantic, and giving many facts bearing upon it. The letter is dated at Frankfort-on-the-Maine, Sept. 14, 1870, but was not published until October 5. The following are some extracts:

"Since the commencement of actual hostilities between Germany and France—that is, from about the first week in August—to the present time, we have had in this part of Germany scarcely a day without rain, generally continuous, and often accompanied with thunder-storms. This phenomenon has called the attention of the German press to the subject, and some valuable historical facts connected therewith have been brought to light; and there appears to be little doubt, judging from the data on hand, that the many storms and rains which we have had in Germany for the past six weeks—a most unusual thing at this season here—have been brought on by the cannonading and firing of small arms in Alsace and Loraine.

*   *   *   *   *   *   *

"The Leipsic *Illustrirte Zeitung* calls to mind a remarkable phenomenon observed in the revolutionary year, 1849. The city of Ofen lies on the banks of the Danube, here running due south. The hill on which the fortress is situated has an elevation of two hundred and thirty-eight feet above that river. It is surrounded on three sides by mountains; on the south by Gernhardsberg, on the southwest by the Adlersberg, on the west by the Schwabenberg (one thousand two hundred feet high), and on the north by the Geisburg (also one thousand two hundred feet high). As the insurgents, at noon, on the 4th of May, 1849, approached the fortress, the latter commenced firing eighty-four guns, eighteen and twenty-four pounders, in order to prevent the besiegers from planting their batteries. Towards evening the cannonading on both sides was furious, and a stable in the fortress was already on fire. The sky, which had been perfectly clear for a number of weeks, became overclouded, and towards midnight a gentle, fine rain fell, the wind being perfectly calm and the shower continuing from one till three.

A clear morning followed. The previous fine weather continued up to the evening of the 17th, when a fearful storm raged, coming as usual from the west. A house on the Schwabenberg was struck by lightning, and the storm ended with a cloud-burst which cost the besieged the lives of a number of horses and men. This storm is supposed to have been produced by a

SIX HOURS BOMBARDMENT OF THE CITY OF PESTH,

by General Hentzi, from Ofen, on the 13th of May, during which engagement six immense mortars of great calibre produced tremendous concussions for a distance of several miles around.

"In the year 1859, an uncommonly violent hail-storm fell in the faces of the Austrians during the battle of Solferino.

"The Germans bring to mind some very interesting American experiences. In 1861, Lewis called attention in Silliman's American Journal to the fact that violent rains and heavy cannonading appeared to stand in intimate connection. He said (I quote the German) : 'In October, 1825, I observed a very plentiful rain immediately after the cannonading which took place in

CELEBRATING THE CONNECTING OF LAKE ERIE WITH THE HUDSON.

I published my observations on this event in the year 1841, expressing the opinion that the firing of heavy guns produces rain in the neighborhood.

AFTER THE FIRST BATTLE IN THE LAST WAR BETWEEN FRANCE, SARDINIA AND AUSTRIA,

there followed such important rains that even small rivers were impassable; and,

DURING THE GREAT BATTLE OF SOLFERINO,

there broke out such a violent storm that the fighting was interrupted. In July, 1861, McClellan's troops, on the upper Potomac, had four separate engagements on four days, and before the close of each violent rains fell.' * * * * *

"The Bohemian campaign of 1866 was accompanied during the whole course by violent rains. After

THE BATTLE OF KŒNIGGRATZ,

violent rain storms hindered the harvest from being properly garnered.

"The letters of the soldiers in the field, in the present war, are full of accounts of 'sleeping on the wet ground,' and complaints of the inclemency of the weather. The 5th of August, following

THE BATTLE OF WEISSENBERG,

was intensely warm. The night of the 6th was rainy, and the morning following

THE BATTLE OF WOERTH,

when the telegrams of victory came, found the streets full of water-pools and the sky overcast with gray, heavy clouds. Since then, we have not had six fine cloudless days. 'From the 6th to the 31st of August,' says the *Illustrirte Zeitung*, 'it rained every day, often accompanied by thunder; and these continuous and violent rains have caused great damage in those districts where the harvest was not in before the 6th; the corn has been washed out, the straw has been rotted, and the crops have no more value.'

"THE BOMBARDMENT OF STRASBURG

is accompanied by the grandest meteorological spectacles. The thunder of the cannon, the blazing of the houses, and the curve fire of the shells, are often intermingled with the roar of thunder and the flashing of lightning. The storms seem to come from the Vosges, to break over the doomed city, and then to spread over the valley from the Rhine to the Schwarzwald, where the grass and trees are almost as green as in spring; and it is well known that, when the war was declared, Baden, Alsace and France were suffering from drought. Great rains fell in Hungary on the 15th of August, the day after

THE FIRST BATTLE AROUND METZ.

In Germany the grapes will be spoiled unless the sun shines with its usual power. We are inclined to think that the storms here are caused by the firing in Alsace, and, up to a week ago, by the bombardment of Strasburg. For the past three or four days, fine weather has set in; and it is a fact that the firing at Strasburg is no longer carried on so strongly, the King having sent orders that the city should now be spared as much as possible from shells. We have had thunder-storms here which surpassed in grandeur and power everything in the experience of the 'oldest inhabitant.'"

The foregoing facts, showing that in different parts of the world, and in all seasons, heavy artillery firing is almost invariably followed by rain, are believed to be sufficient to establish the proposition that to produce this phenomenon at will is within the reach of human power. Some of the storms which have been mentioned would doubtless have occurred if there had been no cannonading; indeed, in such a large number of days, the chances are that, on some, rain, in any case, would have fallen. But no calculation of chances can make it appear a reasonable supposition that rain would have occurred on all, through the ordinary operations of nature alone. The average annual number of thunder-storms, in the latitude of the United States, is a fraction less than 20*—a number totally insufficient to give them with the almost unfailing regularity with which they occurred.

We have seen that, in our late war, almost every battle of the Eastern armies was followed by rain; that rain followed all the great battles of the West, and most, if not all, of the battles of the South; that it not only followed single battles, but fre-

* The average of thunder-storms annually, between latitudes 30° and 50°, is 19 9-10.—Silliman's Philosophy, p. 660.

quently each engagement of a series; that on land and on the water, in the interior and on the coast, on the Mississippi, on the Gulf and on the Atlantic, again and again, storm followed battle; and that the phenomenon, confined to no section, was also peculiar to no season. We have seen that not only in our late war, but in the Mexican war, as well, it occurred again and again, and even in the very midst of the dry season; and that in North America, in South America, in Asia and in Europe, it has occurred under circumstances which compelled the attention of the observers. These facts have a significance that cannot be lightly set aside, and if they do not furnish the positive proof, they fully warrant the belief, that artillery firing always tends to bring rain, and is often the actual cause of its occurrence.

But the question will naturally arise, why is not every battle followed by rain — why is it that an amount of artillery firing that in one case brings rain, will sometimes fail to bring it in another? This question can be better answered after a further consideration of theories as to how cannonading produces rain at all; but it may be remarked here, that it is by no means certain that heavy cannonading does not invariably cause rain somewhere, even if it does not at the spot where the firing occurs.

In twelve instances of heavy firing by the naval and military forces operating against Charleston, that I have investigated, occurring from January

31, to September 5, 1863, ten were followed by rain, and each of the others* by an overclouded sky. In one of the latter cases the sky became overclouded for sixteen hours on the second day after the engagement, and in the other case for twelve hours on the day after; from which facts it may fairly be surmised that it rained in both cases at some place not far distant.

If it be considered a fact, however, that heavy artillery firing sometimes fails to produce rain, either at the spot where the firing occurs or at any place in range of the surface winds, or of air-currents at a higher elevation—still, this would not be evidence that a way may not be discovered to so conduct the firing as to bring rain at all times with unerring certainty. It is probably in the manner of the firing, as well as in the amount, that we must look for the peculiar influence that brings the rain. If it be the heat alone that the firing evolves that exerts that influence, then it would depend principally upon the amount; but I do not believe such to be the case. Professor Espy, some years since, claimed that he could cause rain by means of great fires, and cited rains following battles and conflagrations as facts in support of his theory. He maintained that all rains were caused by an ascending current of air, which drew the air to it from all directions, and that the condensation of the aqueous vapor was due to the cold caused by the expansion of the air as it rose. He believed that all that was necessary, in order to

---

* Attacks of April 7 and Sept. 5. Log of the New Ironsides.

bring on a rain, was to build a great fire so as to heat the air and cause it to ascend, and that after an ascending column was once formed, the fire might be allowed to go out, and the upward current would still continue. The manner in which the process would go on—and, according to his theory, the process is the same in the case of any rain—may be thus described: The air, as it rises, expands; by its expansion it becomes cold; by its cold a portion of its aqueous vapor is condensed into cloud; by the latent heat given out in this condensation, the air is warmed and made to rise higher, causing a further expansion and a further condensation of vapor; the surrounding air, in the meantime, rushes in to take the place of that which is thus ascending and passing off, and goes through the same process.

But, though Mr. Espy's theory contains much that is true, he failed, as a general thing, to convince the philosophers of his day that rain is produced by such a process, and in their opinion the writer concurs. One objection that I have to the theory is, that it seems to teach, in substance, a contradiction—that the air becomes cold by expansion, and warm at the same time by the effect of that expansion, without an equilibrium being produced, and with it an immediate stoppage of the process. Another is, that it teaches that the mixture of air and aqueous vapor contains within itself the power to rise, and that the upward current, after being once started, forms what would be, with all circumstances favorable, a sort of perpetual motion, thus seemingly ignoring the fact that it requires as much power to lift up from the earth to the region of the clouds any definite amount of water, in the form of an invisible vapor, as would be required to lift it up in a solid body. Heat is the power that raises to the upper regions of the air the vapor that forms the rain, and this heat comes from the sun; and the rise of the vapor to the requisite height is probably gradual, and fully effected before the occurrence of the rain.

The heat given out by the burning of gunpowder in a battle undoubtedly causes an ascent of air, and with it of vapor; but this effect must be limited. It is manifestly absurd to suppose that the power residing in the heat evolved at such a time is sufficient to lift up from near the earth to the region of the clouds the vast quantities of water which, so often, afterwards come down.

I believe that the rain which falls after a battle is drawn from vapor which, having previously risen by the heat of the sun, exists at the time at a great height, either in currents of air flowing from over the sea or from the tropics, or in comparatively still air containing vapor, some of which has risen from the earth below, and some of which has been left by air-currents such as those above referred to.

I have suggested that electricity may be an active agent, as well as a result, of the condensation of the invisible vapor of the atmosphere into clouds and rain. If this conjecture is correct, we can in part understand why cannonading brings rain, by supposing that one of the primary effects of the cannonading is to develop electricity in the air, or cause some peculiar form of electric action. Electricity is regarded as one of the forms of force; and certainly, in the discharge of a battery of artillery, an immense force is brought into action. Friction is one of the means by which force is converted into, and made to produce electricity; and if two batteries of artillery are placed opposite to each other, and at a suitable distance apart, and fired simultaneously, can it be doubted that friction will be produced by the particles of the air moving over each other, and that electricity will be developed?

It is frequently the case that the first phenomenon following a battle is a dense fog. I have given but little prominence to this circumstance in the foregoing statement of facts; but it would seem to show that concussion, by producing electrical action, or in some other unknown way, may operate directly to cause condensation of aqueous vapor. But it is likely that, in the actual accomplishment of rain, a more complicated process takes place than in the production of fog, and in this process motion is probably one of the chief factors. If, as Hutton's theory maintains, the condensation of vapor into clouds is caused by the mixing of two bodies of air of different temperature, motion is requisite to effect such a union. Motion is also, in any case, necessary for an accumulation of vapor in sufficient quantities to make any considerable amount of rain. To produce clouds dense enough, the air which supplies the vapor must deposit it within a much smaller space than itself occupies, which can only be effected by such a motion as will bring different portions of it in succession to the place where the clouds are forming, and carry it off again as fast as each precipitates its vapor. Motion is also necessary to bring the watery particles into contact with each other so as to collect them into drops of rain. There are two ways in which concussion may be supposed to communicate this motion; one is by causing a change in the motion of air-currents already existing at a high elevation. The violent effects of heavy concussions in other ways are too well known to need more than a passing notice. Indeed, so terrible have they been in some cases, that at the first attack on Fort Fisher, the experiment was tried of exploding a ship-load of gunpowder before the walls, in the expectation that the shock would so paralyze the garrison as to render easy the seizure of the fort. Though this expectation was not realized, the failure was probably owing less to an erroneous idea of the principle involved than to the manner of its application. The explosion was probably made too far from the walls of the fort, and it may have been that the powder was, much of it, stowed away in the vessel below the waterline, which would cause much of the force of the explosion to be in an upward, instead of a lateral direction. An instance of the effect of the concussion of heavy artillery firing was seen at the bombardment of Forts Jackson and St. Philip, in the numbers of dead fish* that floated down the river; and it has been noticed that buzzards give a wide berth to a region in which there are frequent military conflicts,† which they would scarcely do, unless the effect of the concussion extended to a great distance, and were at times decidedly uncomfortable to them, probably most so when they are on the wing. The breaking of windows by a discharge of cannon is a not unusual occurrence; and the statement that water-spouts at sea are broken up by the same means is one with which all are familiar.

These considerations, and especially the latter, suggest the idea that the effect of artillery firing upon a horizontal current, or moving stratum of air, may be very great—the greatest effect being directly above the place where the firing takes place. Such a current, being necessarily in exact equilibrium with the air above and below it, would yield to the slightest force; and it may easily be conceived that the concussion of a number of cannon, fired simultaneously, would first cause it to sway upward, and finally, if the firing were continued, to break up. This being accomplished, we should have a condition of things which, according to accepted theories, might bring rain—we should have two bodies of air of different temperatures, and different degrees of humidity, becoming mixed together. If the current of air were warm and charged with

* Am. Conflict, Vol. II. p. 89.
† See Doc. No. 19.

vapor, then the drier air above it would be colder
—this condition being necessary to maintain the
equilibrium which permitted the existence of such
a current. If, on the other hand, the current
of air were colder than that above it, then the latter
would contain more vapor, and it is possible that it
might contain enough to produce rain.

But more favorable conditions may be supposed
to exist than either of these, to give full effect to
the action of concussion. Let us suppose two currents of air, one above the other, and moving in
a different direction, and that one is cold and the
other warm and saturated with vapor. The effect
of heavy concussions would probably be to throw
them upward over the spot where the firing takes
place, and finally, with their own motion, to bring
them together. A spiral motion would result from
their union; larger and larger portions of the two
currents would become involved in the change; the
two bodies of air would become mixed; and the
conditions required by the Huttonian theory of rain
would be fulfilled.

The objection might be made to the idea that
concussion would throw an air-current upward in
the way I have conjectured, that it would cause a
rarification of the air below it; but the difficulty is
not, I think, a serious one. The gases given out by
the burning of the powder would help to fill up the
vacuum which the projection upward of the current
would tend to create; the heat given out, by expanding the air, would further assist in this direction,

and the equilibrium might be further restored by air thrown off from the current itself, or by a slight dropping down of the surrounding portions of it not greatly affected by the concussion. This, also, would explain why the first motion of the barometric column in a battle is generally, or at least frequently, a slight rise — as the amount of air above the place of the firing would be increased by the above process.

But another, and perhaps more serious objection, would be, that the force of concussion would be so diminished at a great height that it could not produce such an effect on air-currents as I have conjectured. On this point it must be admitted that there is room for a difference of opinion, and that more precise knowledge is needed; but, in addition to what has been said in relation to the force of the movement which we call concussion, as exhibited in other ways, and in particular in its effect upon a water-spout at sea, its action upon smoke under certain circumstances may also be cited as showing that it must have a peculiar effect upon air in motion. In this connection I would call attention to what is said in the interesting letter of Gen. McNulta, which will be found on another page, in relation to the effect of the firing at the inauguration of Governor Hahn, at New Orleans, upon the smoke issuing from the chimneys of the city; a kind of pressure seemed to be exerted, very much greater than would be supposed.

Another process by which concussion may be supposed to give motion to air remains to be considered. Returning to my first supposition, that concussion has a direct effect in causing a condensation of aqueous vapor, let us suppose that it acts upon air at a great height which is comparatively without motion at the time. Such air may not unreasonably be supposed to contain sufficient vapor to produce rain. Professor Espy has shown that when air saturated with vapor rises, portions of the vapor ought to become condensed by the cold of expansion; but facts of every-day occurrence show that this is not what always takes place. If it were so there would be constant rain at the tropics, where there is a great evaporation of water; whereas, instead of the vapor falling back as rain, it must, to a great extent, pass off northward and southward in the equatorial currents. Hence, at a great height there may exist large quantities of vapor, and at the same time a great degree of cold without the latter of itself effecting condensation. Concussion may supply the conditions which are wanting. Upon the condensation of aqueous vapor into cloud, the latent heat of the vapor would be given out, which, warming the air in which the condensation takes place, would cause it to ascend; the surrounding air would rush in to take its place, and a motion of the nature of a whirlwind would result, which would be of greater or less extent, according to the amount of air acted on and the intensity of the action.

It is known that there is a current of air flowing at a high elevation, from the equator towards the poles; and it may be that in this, or from vapor which it in part supplies, are generated the storms which follow battles. If this be a reasonable conjecture, we may suppose that there is an unfailing source from which to draw rain at any time, excepting, perhaps, immediately after we have brought on a heavy rain—in which case we might be obliged to wait until the air which furnished it had passed off before we could produce another.*

The foregoing conjectures as to the manner in which artillery firing operates to bring rain, may be summed up as follows: That the effect is due principally to concussion, and that the operation of concussion is two-fold, producing electrical action, and change or communication of motion. The fact that we cannot see a spiral motion in the clouds does not show that there is not such a motion in the air immediately above the clouds. When they commence forming they probably also commence falling to a lower and denser stratum of the atmosphere, and this hides from our sight the process that is going on. Variable winds and sudden squalls frequently follow battles, which may perhaps be due to the effect upon the lower air of a whirling motion above.

Volcanic eruptions and great conflagrations probably effect the same results as cannonading, in a different way—heat being the agent instead of concussion. In these cases, the disturbance of the upper air is probably occasioned, to a great extent,

* See remarks in Doc. No. 27 bearing on this subject.

by the upward motion of heated air — a motion which must be far greater than during a battle, owing to the greater amount of heat evolved.

In attempting to explain why artillery firing does not always bring rain, I will remark, first, that to produce the above described effects in the greatest degree, it is probable that a certain arrangement of the guns is necessary, and that they should be fired simultaneously. The greatest electrical effect would probably be produced by arranging the guns in two lines, a mile or two apart, and pointing them towards each other, with a little elevation. The greatest effect in directly communicating motion would probably be obtained by arranging them in a circle, or massing them, with the muzzles pointing towards the zenith. A study of the movements of the barometric column during battles, as recorded in the log-books of the navy, has led me also to conjecture that two spells of firing, a few hours apart, will produce a greater effect than would the same amount of firing commenced and finished without an intermission.

In a great battle, there is such a diversity in the arrangement of the guns, and such a large amount of firing, that it generally happens that all the conditions necessary to produce rain in very large quantities are fulfilled. Yet it may be that it is only a small portion of the firing that has an actual effect in accomplishing this result. It is probable that it is only the simultaneous explosions from a number of pieces of artillery that have any power in that direction, and it may be also that a number of pieces fired at the same instant will have but little effect, if they are all pointed the same way. If electricity is developed in the manner conjectured, it is necessary that there should be a meeting of opposing shocks of concussion and waves of sound. The same thing is also necessary in order to produce the greatest motion upward when the guns are pointed with only a slight elevation, as in battle. It may be also that, except in very favorable circumstances, and when the first firing is exceedingly sharp and concentrated, it is requisite that after a few hours a second impulse should be given, by a second spell of firing. These considerations, I think, will explain why some instances of heavy firing have not been followed by rain. The amount of the firing may have been more than sufficient, and the fault may have been wholly in the manner of it.

The question, however, as to whether rain can at all times be produced, and whether or not it can be made to pay, can only be settled by experiments. These experiments ought to be conducted near arsenals, where there is an ample supply of cannon, so that the number and weight of ordnance that could be most economically used to produce the greatest effect could be determined. In conducting the experiments, the first and most obvious arrangement for the guns would be to place them in two lines, some distance apart, the guns in each line pointing towards the other. They should be connected by an insulated wire, and all fired at once by means of electricity.

Another experiment would be to arrange the guns in a circle of a mile or two in diameter, pointing their muzzles toward the centre and upward, and firing them simultaneously, as before.

A third experiment would consist in massing the guns and firing them simultaneously with their muzzles pointed towards the zenith.

A variation of these different arrangements of the guns will suggest themselves, and need not be here adverted to. Another important point to be determined, as well as the proper number and arrangement of the guns, would be the time that ought to elapse between the different explosions. I am disposed to think that a spell of rapid firing, followed in a few hours by another a little heavier, would produce the greatest effect with the least amount of powder. This idea is suggested from noticing that at some of our naval battles it did not rain until the second fall of the barometer. The movements of the barometric column during battles, however, are frequently so irregular—owing partly, perhaps, to other causes than the firing—that it is difficult, from the recorded observations so far made, to fix a definite law for them. I think, however, that it will be found that an experiment of heavy firing, performed in a state of the weather when there are no winds or other disturbing influences, and when the barometric column is perfectly quiescent, will be followed, immediately, by a slight rise of the mercury; that in a few hours the latter will have fallen lower than it stood before the firing;

that the fall will be succeeded by another rise, and that if, while this is taking place, the firing be renewed, the rise will be followed by another fall, and with it a fall of rain. There are indications in the log books of the navy, that the barometer will sometimes fluctuate, even if the first spell of firing is heavy enough to bring the rain; and there have been cases where the sky became overcast, then cleared off, and then became overcast a second time before it rained, and all within a short space of time. At the attack on Fort Fisher, where the powder-boat "Louisiana" was exploded, there were several fluctuations of the barometric column, rain appearing at its fourth depression, counting from the time of the explosion of the boat. The commencement of the bombardment was followed by an immediate rise of the barometer, though this was not the case after the explosion of the boat. On board the steamer "Malvern," the barometric height at 1.30 A.M. was 30.26 in. Sixteen minutes later, the officer of the watch saw the flash of the explosion and heard the report; he examined his barometer and found the height to be 30.24 in. At this height it stood until 4 A.M. At 6 A.M. it had fallen 0.06 inches, the height then being 30.20 in. At 8 A.M. it was 30.30 in., and the same at 9. But from 9 to 10 A.M. it fell 0.19 in., standing at the latter hour 30.11 in., at which height it remained at 11 A.M. At half-past 11 the bombardment of the fort commenced, and at 12 the column had risen 0.21 in. standing at this hour 30.32 in. During the second depression of the barometer, after this, it rained; the height at the commencement of the rain being 30.10 inches, from which it fell during the rain two-tenths of an inch, rising again to the same height before its close. During the second rain—the one that took place at noon on the day following the second day of the bombardment—the barometer stood at 29.84 in.

That these fluctuations of the barometer were not caused by wind, the following considerations will show: Commodore Thornton A. Jenkins, late Chief of the Bureau of Navigation, Navy Department, in a valuable work written by him, on "the Barometer, Thermometer, Hygrometer, and Atmospheric Appearances at sea and on land," says, in regard to the direction of the wind, as affecting a barometer, that the northeast wind tends to raise it most, and the southwest wind to lower it most, and winds from points of the compass between them proportionally as they are nearer one or the other extreme point (page 11). Now, when the powder-boat exploded, the wind stood W. by S. At 6 A.M. the wind was W. N. W., and the barometer, according to the above rule, should have risen a little, whereas it had fallen. At 9 A.M. the wind was West, and the barometer ought, if anything, to have fallen from its last height, instead of which it had risen. At 10 A.M. the wind was W. S. W., which was nearly what it was at the time of the explosion. Again, at 11 A.M. the wind was S. W. by W., and at 12 it was S.W. If this slight change of the wind had caused any change of the barometer, it ought to have caused a fall—instead of which there was a rise of 0.21 in., the bombardment having in the meantime commenced. Again, during some of the other shifts of the wind, it was for a time from the northeast, in which quarter the barometer went 0.22 inches lower than it stood when the wind was from the S. W., showing the rule again reversed as regards the effect of the direction of the wind alone upon the height of the barometric column. During one of the greatest of these fluctuations there was no change in the force of the wind.

The change in the height of the barometer, caused by an ordinary naval battle, is not generally very great, seldom much exceeding two-tenths of an inch. At the bombardment and passage of the Vicksburg batteries by Admiral Farragut, on the morning of June 28, 1862, the movement was much less even than this. The barometer at midnight— three hours before the commencement of the engagement — stood 30.00 in., with no wind; and at noon, after the engagement, it had risen to 30.03 in., with wind variable and very gentle. After this it commenced to fall, and at 4 o'clock the next morning it was 29.98 in., with rain and with wind from N. W. The rain after this battle appears to have occurred during the first depression of the mercury. There had been slow firing by the mortar-vessels the day before, however, with first a rise of the barometer of 0.03 in., and then a fall of 0.04 in.*

After the naval action off Charleston, Jan. 31,

* Log of the U. S. Steam Sloop Hartford.

1863, there was a fall of the barometer of nearly half an inch.

It will not be necessary, in order to procure rain, that there should be a cloud in sight when the firing is commenced, nor that the wind should blow from any particular quarter. To obtain the greatest amount of rain, at the place of the cannonading, it is probable that a calm state of the air will be most favorable; if rain is wanted at some other place, however, the direction and force of the wind will need to be taken into consideration.

While the manner of firing will probably be of so much importance that an amount, which, in one case, will bring rain, will fail to do so in another where it is differently conducted; yet upon the amount of firing, where the manner is the same, will probably depend, in a great measure, the amount of rain produced. Great battles are generally followed by greater rains than are minor engagements; and if the theory which I have advanced, as to the effect of concussion upon air-currents is correct, this would necessarily follow: as the greater and more numerous the concussions, the more extensive would be the change of motion which would be communicated, and the greater, within certain limits, would be the volumes of air of different temperatures that would be mixed and caused to deposit their vapor.

A question pertinent to the subject is, whether we can stop a rain, as well as bring one on. Probably most of those who have given thought to the subject would answer in the negative, if the rain has been brought on by cannonading; but the remedy here will be to so regulate our firing as to bring only what rain is needed. Neither is it likely that human power can ever exercise any control over great storms, such as sometimes sweep over the country. The best that can be expected to be done with these, is what the signal service, by its system of weather telegrams, and its proposed storm signals, has undertaken, to give warning of their approach. But local rains, extending over a limited area of country, may, perhaps, be successfully dealt with, especially if it can be determined from which quarter come the currents of humid air that keep them up. In this connection the following fact may be mentioned, though it may not, perhaps, be of much significance. Off Pensacola Light, on the morning of April 12, 1861, the weather was stormy and ugly up to 4 o'clock, but after the bombardment of Fort Sumter, at Charleston, commenced, it changed and the storm passed off.* If the question should ever be decided by experiments, it will, very likely, be found that to put a stop to a continued succession of showers, happening in one place, it may be necessary to bring one on somewhere else.

Judging from the letters which I have received since commencing an attempt to bring this subject forward, I believe that the country would regard with interest some experiments in the matter,

* Log of the U. S. Frigate Sabine.

and would not begrudge the expense, even if they should prove unsuccessful in leading to a practical use of the principle under discussion. In some other matters connected with science, the government has justly considered that an expenditure of public funds was calculated to be of public benefit; but where, if we except in the system of weather telegrams and storm signals that it has recently undertaken, was there ever so promising a field for such action as here. A just and equal regard for the interests of different classes of the people also requires that, if the production of rain at will and at moderate expense is within our reach, it should be known, and known, that it should be acted on. The system of storm telegraphy is for the benefit of commerce; let the interests of agriculture also be considered. If it is a legitimate subject for legislative expenditure, to provide means for giving warning to the merchant and shipper of the approach of hurricanes, it is no less so to provide relief for the farmer when his fields are parched with drouth, if this is practicable. A remark of Prof. Maury, in writing on the former subject,* will apply with equal force to this; he says: "Hundreds of thousands of dollars are lavished upon scientific expeditions for the observation of eclipses, for prosecuting geologic speculations, the survey of distant lands, and even for explorations in Arctic ice in search of the mysterious pole. How insignificant are such objects when placed by the side of that now before us!"

Scribner's Monthly, Feb. and March, 1871.

I do not propose that the government should establish stations through the country, and proceed to furnish rain in different sections as it is needed; far from it. What is known of the subject is entirely insufficient, as yet, to warrant such a procedure. When the power of steam was first discovered, the world was not ready to build steamships and railroads; nor, when the first electric battery was made, was it ready to lay telegraphic cables across the Atlantic. But from small beginnings, that promised much less than this, how much has the world accomplished! Yet it has only been done through experiments, patient and persistent; experiments that, had they been as costly as those which are now proposed, would never have been made, and man would have remained to this day unconscious of half his powers.

But the proposed experiments, though costly, considered as an individual undertaking, would be but a trifle to a great nation like ours. We have the powder, and we have the guns and the men to serve them, and we ought not to leave to other nations, or to after ages, the task of solving the great question as to whether the control of the weather is not, to a useful extent, within the reach of man.

I append some letters from distinguished officers and others in regard to the matter under discussion. My apology to the writers, for the use so made of their favors, is found in the importance which I believe the subject to possess, and in the impossibility of presenting it in its proper light, except by showing how the phenomena which I have described are regarded by some of those who witnessed them. As documents in support of what has been advanced, they are too valuable to remain hidden from the world; and, as they relate wholly to matters connected with history and natural philosophy, I believe there can be no impropriety in making them public.

APPENDIX.

APPENDED DOCUMENTS.

Doc. No. 1.

From Brevet Major General Elisha G. Marshall, U. S. A.

ROCHESTER, N. Y., *Dec.* 7, 1870.

Mr. EDWARD POWERS:

Sir — Yours, of Nov. 28, was received a few days ago. In reply, I would state that your article I have seen in many of our scientific papers, and was then pleased with your views, as often myself and other officers, during the war, were well acquainted with the fact that artillery firing, etc., caused rain.

I will give you facts, which I happen to recollect as far as Grant's campaigns.

First Bull Run — One day's fight. Heavy rain next day.

Second Bull Run — Two days' fighting. Heavy rain day after fighting, extending beyond our retreat at Centreville.

Malvern Hill — Two days' fighting. Very heavy rain next day after battle, extending to our retreat at Harrison's Landing.

First Fredericksburg — Heavy rain after fight.

Antietam — Heavy rain.

There was a rain after Chancellorsville, and, as far as I can recall, after every battle where much artillery was used through all of Grant's campaign.

Grant's campaign was more of one continuous fight from Wilderness to end of war, so that I would not pretend giving data.

The above notes are given you after conferring with Brevet Major General C. J. Powers, Vol., Col. 108 N. Y. Vols., who happened in at this time. Your theory, in reference to this phenomena, I consider correct, and deserves full consideration from Congress, and research; and I shall be glad to assist you in obtaining the hearing of those you wish, as far as my humble means go. You will find that every officer, almost, of any education or thought, will be apt to agree with your views, as this matter was often spoken of during the war.

Is it not the same principle we call to operation when we fire artillery over the spot of a drowned person?

Cannot the Surgeon General, from his surgeons' meteorological observations, give you fuller data,* or put you in correspondence with his corps, who were present at every battle, and they, after careful thought, give you all you seek? You will find the reports of army surgeons reliable.

Truly,

E. G. MARSHALL, U. S. A.

Doc. No. 2.

From General Joshua L. Chamberlain, Governor of the State of Maine.

STATE OF MAINE, EXECUTIVE DEPARTMENT,
AUGUSTA, *Dec.* 12, 1870.

MY DEAR SIR — My Adjutant General has sent me your letter, referring to the effect of heavy firing on the atmosphere leading to storms and rain. It is a most interesting matter. The *fact* of such sequences (if they may be called so, without begging the question,) I have often noticed. Certainly a heavy storm of rain occurred after the great battles of Antietam, Fredericksburg, Chancellorsville, Gettysburg, the Wilderness, Spottsylvania, Bethesda Church (or Coal Harbor), Petersburg, Five Forks, etc.; and often, I well remember, in what we called small engagements (though they would be called battles in Europe,) such as the fight on the "Quaker Road," March 29, 1865, for a late instance, in which there was a sharp, concentrated fire of infantry and artillery for a couple of hours, a very heavy rain would surely

---

* Col. C. H. Crane, Assistant Surgeon General, says, in reference to this matter: "Very few meteorological reports were sent to this office during the war, and those few came from posts distant from the scene of hostilities."

follow. This fact was well noticed, and is well remembered by many a poor fellow, who, like myself, has been left lying, desperately wounded, after such engagements—for these rains are balm to the fever and anguish of the poor body that is promoted to the ranks of "casualties."

I am sure you will find my testimony confirmed by the recollections of every soldier.

* * * * *

JOSHUA L. CHAMBERLAIN, *Governor*,
Late Brev. Maj. Gen. commanding 1st Div., 5th Corps.

Doc. No. 3.

From General Elliott W. Rice, late of Iowa.

OFFICE OF ELLIOTT W. RICE, ATTORNEY AT LAW, }
1424 F ST., WASHINGTON, D. C., *Nov.* 3, 1870. }

Mr. EDWARD POWERS, Chicago, Ill. :

My Dear Sir — I have your letter of Oct. 22, enclosing your letter in the *Post*, in relation to storms, produced by firing of cannon. I remember well that many of our heavy battles were followed by rain. At Donelson the weather was clear and cool, and exceedingly pleasant; but, soon after the engagement, a snow storm was upon us, which was followed by rain. Sunday morning of the battle of Shiloh was clear and beautiful, almost—yes, entirely—beyond description. The day's terrific battle was followed by a drenching rain, that all, who were there, must well remember. The same thing occurred in the Atlanta campaign, particularly at Dallas; also, at first Bull Run; at Gettysburg. In fact, the occurrence was so frequent, that there can be but little doubt that the rain, in many instances, must have been produced by the commotion produced by battle. A Confederate Colonel, now in my office, informs me that it was frequently remarked, in their army, that great battles were frequently, if not generally, followed by storm. I trust you will pursue this interesting subject, which may result in a discovery of incalculable benefit to the world. I regret that I have not time to write to you more fully on the subject.

Very truly,

E. W. RICE.

Doc. No. 4.

From Gen. John McNulta, of Illinois.

BLOOMINGTON, ILL., *Dec.* 13, 1870.

EDW. POWERS:

My Dear Sir — In reply to your favor of the 10th, not only has it been my experience that rain follows soon after every heavy cannonading, but that this was very generally conceded and understood in the army, and acted upon by the soldiers in preparing for it after every battle I remember, particularly, that in the garrison at Lexington, Mo., when water could not be had, it was urged by myself and other officers encouraging the men to hold out for a few hours, and that the cannonading would bring rain to quench their thirst; and it did bring the rain, but found us without the means to catch it in sufficient quantities. There are large numbers of soldiers in your city who will remember this circumstance, and the wringing of their blankets to get water.

I have often thought of the matter, and am well satisfied that the theory you advance, in the printed slip sent me, is correct; and, also, that the rain is produced quicker when there is no wind (unless it be a wet wind,) than with wind, unless there be a range of hills or mountains to the leeward.

The officers of the Mississippi fleet could, I think, give some important facts, from their shelling small squads of the enemy on shore, with reference to this matter. The inauguration of Gov. Hahn, at New Orleans, was accompanied with cannonading, and noise of musical instruments and anvils, infantry firing, etc., and was soon followed by very heavy rain. When the rebel ram ran by the city, the cannonading, only for a few minutes, was followed by rain. The passage of the forts at Mobile bay, the bombardment of Fort Gaines, afterwards of Fort Morgan; again of Spanish Fort and Blakely; the landing of our troops at Pascagoula, and firing a few shots with field pieces on shore; the battle of Sterling Farm, and the fight on the Atchafalaya river were followed, in a few hours, with heavy rains.

I was with the first troops that passed down the river (Herron's Division, 13th Army Corps,) after the surrender of Vicksburg, to Port Hudson. We found it very muddy there (July,) and also at Yazoo City, when taken by our troops, July 12, 1863. Everybody remembers that there was no trouble in keeping *moist* at Vicksburg. It rained, after cannonading, at all the places named; but why I have named this region of country is, because it seemed to impress me with its peculiar susceptibility, in this respect. I believe that in the dryest time, without wind, or a light wind off Ponchartrain, the firing of one hundred guns at New Orleans will bring on rain in a few hours, and almost certainly in large quantity.

At the inauguration, of which I speak, I obtained, during the firing, a seat on a house-top, on Lafayette square, where I could look down and see the multitude. It was, literally, as *dry as a chip*, without a cloud to be seen, when the ceremonies commenced. (Chimneys presented the appearance of miniature volcanoes, spasmodically sending up soot, and here and there one with fire and ashes. The escape of the compressed air from a chimney, occasioned by the concussion of artillery, is infinitely greater than would be supposed. There seems to be a sort of hydraulic pressure to it.) My recollection now is that the artillery practice had not commenced more than an hour when there was a perceptible change in the atmosphere — a kind of haziness. That night, and the next morning, it rained "*fearfully* hard."

What effect would be produced on our prairies, I am unable to say; yet believe the difference could not be great, as the aqueous vapors contained in our atmosphere cannot be much less than in that near the ocean or large rivers. At the battle of Prairie Grove, Ark., the wind blew parallel with the mountain range. I suppose there was an average of twenty field guns in constant use for five hours. Cloudy at dark or a little after, (Dec. 7.) At 2 o'clock, A. M., 8th, the atmosphere was remarkably clear, and the stars shone with unusual brilliancy. At 3:30, it was "pitch dark." Daylight showed a few filmy clouds, with the light wind blowing against the mountain range. M. brought rain.

Soon (several days) after we crossed the Boston Mountains, we found a light breeze blowing against the mountains from the opposite side. We had some artillery firing — say thirty or forty rounds — near Lee's creek, early in the morning, with a clear sky. Here I remember that it was urged, by some of our officers, that artillery should not be used on the small number of the enemy's cavalry that were in front of us, for the reason that *it would bring on rain*, and thereby retard us in the pursuit of the enemy. We got the rain in less than two hours. After the firing, at the capture of Van Buren, the wind still light, moving nearly at right angles with the mountain range, we got rain in a few hours.

I remember that the rain following the several engagements on Mobile bay, was more copious than I ever witnessed before — the cannonading by the army and navy being unusually heavy, especially at the lower forts. This, you will remember, was in the summer, or dry season. It is possible that I may have known an instance where there was heavy firing that was not followed by rain, and that the matter may have escaped my notice. There are other instances when I know it did occur, but deem it unnecessary to cite them, as I have given you the cases that seem most important, and which attracted my attention. I am compelled to hastily refer to the matter, and have scarcely time to read what I have written, and would not have written this much but from a desire to encourage you in prosecuting your investigations, for I seem to know you are right. I hope you may succeed in fully demonstrating your theory, and in making some practical use of it.

There are, as you doubtless are aware, animals that anticipate a storm — hogs, for instance. If you have an opportunity, after the firing of a salute of say thirteen or fifteen guns, observe them carefully, and I think you will find them acting as they do preceding a storm, although there may be no rain.

Truly, yours, J. McNULTA.

Doc. No. 5.

From Gen. R. H. Milroy, of Indiana.

DELPHI, IND., *Dec.* 19, 1870.

Mr. EDWARD POWERS:

Dear Sir — Yours, of the 8th inst., was duly received, containing the enclosed printed slip, and asking me to state, whether, in my military experience, I had noticed the fact "that battles are generally followed by a rainfall." I regret that my attention was not called to this matter during the war that I might have noticed and made note of such facts, as I doubt not their existence. The Espy theory of producing rain is the only one I recollect hearing of prior to our late war. About the close of the war I heard it mentioned that heavy artillery firing produced rain, and in looking back over battles in which I participated, or was near, I thought I could see strong proof of the theory. The battle of Rich Mountain, in July, 1861, was followed by some one or two rainy days; Cross Keys, in June, 1862, the same; second Bull Run, August, 1862, two days' heavy firing, in dry weather, followed by refreshing showers; Gettysburg battle, heavy artillery firing for two days, in July, 1863, followed by such heavy rains as to raise the Potomac, and stop the retreat of the rebel army for some days; battle of Franklin, Tenn., about the 1st of December, 1864—fine weather at the time of and previous thereto for many weeks, but was followed by rain that froze as it fell, and covered the country with ice; battle of the Cedars, near Murfreesboro, Tenn., in December, 1864, preceded for some days by artillery firing from Fortress Rosecrans, was followed by rains; the battle of Nashville, soon afterwards, was followed by much rain.

These are all the instances I can now recall of " Heavens weeping o'er our battles." There is more artillery being used in the battles now going on in France than was ever used in any preceding war on earth, and the newspapers tell us that France has been having extraordinary quantities of rain and great floods. May not this fact account for our pleasant dry fall and winter, so far?

There are none of the laws of nature of which science is so utterly ignorant as those governing weather, and yet there are no laws the knowledge of which would be of more benefit to mankind. You are, therefore, engaged in a most noble and beneficent inquiry, and I most sincerely wish you success, and hope that Congress will grant you all you ask to enable you to prosecute your experiments. There is no reason why science should not obtain a knowledge of and utilize the laws governing weather.

Begging pardon for my delay in answering your inquiry — occasioned by my press of business — and for the hasty and unstudied manner of the foregoing answer,

I am, very respectfully and truly,
Your most ob't serv't,
R. H. MILROY.

A distinguished officer, who takes an adverse view of the practicability of producing rain in a time of drouth, says: "The difficulty is this: to cause rain by concussion of the atmosphere, you must have the atmosphere charged with aqueous matter — a thing beyond your control. Therefore, while I believe that when charged with moisture, violent and protracted concussion may precipitate and hasten the fall of rain, I doubt whether, in the absence of a proper hygrometrical condition, (which is always the case in times of drouth,) any concussion would produce rain."

The theory which I have advanced as to the manner in which cannonading produces rain, fully meets the above, by showing how the phenomena may be supposed to occur without being dependent, in the least degree, upon the condition of the air near the earth's surface. But to meet the objection on its own ground, I remark that the less water there is on the earth at any time, the more there must be in the air above the earth; and that the existence of drouth, instead of showing an absence of vapor, indicates rather an absence of the conditions favorable to its condensation — conditions which concussion may produce. That there is no reason to despair of being able to produce rain at such a time, from any supposed absence of a sufficiency of aqueous vapor, is shown from the following letter from one of the most distinguished scientists of the age — a document to which I attach a special value, not only for the light which

APPENDIX. 107

it throws on the particular point under discussion, but for the support it gives to the proposition I have made for a series of experiments.

Doc. No. 6.

From Prof. Benj. Silliman, of Yale College.

NEW HAVEN, *Nov.* 19, 1870.

ED. POWERS, Esq.:

Dear Sir — In reply to yours of the 11th, received to-day, I have to say that it by no means follows that in times of drouth the atmosphere does not contain a considerable quantity of water, dissolved as vapor. The capacity of the air for moisture increases with the temperature, and, in our country, east of the Rocky Mountains, we seldom see a state of the air, where it does not contain a large amount of moisture. The amount of moisture requisite for saturation of air at different temperatures, is as follows, viz.:

1 cubic meter of air, at 0° C, holds 5.4 grammes.
1   "   "   "   " 10° "   " 9.74   "
1   "   "   "   " 25° "   " 22.5   "

In very dry climates the air is often very low in moisture, as at the Red Sea, during a simoon, when not over $\frac{1}{15}$th of satution is present. In this latitude, 60 per cent. of the saturation is a usual and healthful quantity. The "dew point," of course, is the test of saturation. I have seen, in Arizona, 40° F. difference between the wet bulb and the dry bulb thermometer, and there, I believe, no cannonading would bring rain out of the air.

I consider, however, that the matter you have in hand is a perfectly fair subject of experiment, and in view of the fact that there are times (and we have all seen such,) when a good shower would be worth millions of dollars in money, it is certainly worth a few thousands spent in noise, at a proper time, to determine the question, " Will He bow His heavens and come down ?"

Yours, respectfully,

B. SILLIMAN.

As a further answer to the objection that has been referred to, I would call especial attention to the following letter. When, in these United States, is there ever such an "absence of a proper hygrometrical condition," as in Mexico, in the very midst of the dry season?

Doc. No. 7.

From Brevet Major General Henry W. Benham, U. S. Engineers.

BOSTON, Mass., *Nov.* 15, 1870.

Mr. EDWARD POWERS, Chicago, Ill.:

Dear Sir — I have been greatly hurried, during the past few weeks, so that I have not had time to reply, as I would desire to, on the subject of your inquiry, as to my opinions or experiences in the matter of the effect of cannon firing to cause rain at any or all states of the atmosphere.

I would say to you now, however, very briefly, that I have a most decided conviction on this subject — as I have had for many years, — and that is, that the firing of cannon, to any great extent, will always, or almost always, cause rain. Independent of several cases in the last war of the rebellion, where rain accompanied or followed the battles in quick succession, I will only now refer particularly to one case, which, I doubt not, the recollections of many men in your vicinity who were members of Hardin's or Bissell's regiments, at the battle of Buena Vista, Feb. 23, 1847, will corroborate This is the fact: that about one or two hours after the severe cannonading between 8 and 10 A. M. — that is, between 11 and 12 o'clock — we had a most violent rain-fall for some ten or fifteen minutes. I recollect holding my body forward over my holsters, and bringing up my frock coat skirts to keep my holsters and pistols dry. Again, in the afternoon, at about the same interval, after the last fatal charge, when Colonels Yell, Hardin. McKee, and Lieut. Col. Clay fell, — when there was a heavy cannonading a second time, — another violent shower of rain fell, wetting us all again. And what I considered the *satisfactory proof* that this was caused by the shocks to the atmosphere produced by the cannon fire, is the fact that no rain had fallen in that vicinity for many months previously—I was told six or eight months,—and none fell, *as I know* was the case, for three or four months after that battle, as I continued at that position.

Trusting this may be of some use as an item towards substantiating your views, which, I do not doubt, can be utilized, as you proposed,
I am, very truly, yours,
H. W. BENHAM.

If further proof were necessary, that rain can be produced by cannonading, in a time of drouth, it is furnished in the following, taken in connection with what is said in Greeley's History of the American Conflict, Vol. II. page 218, in relation to the weather preceding the battle of Perryville, or Chaplin's Creek.

Doc. No. 8.

From General Geo. W. Smith, of Illinois.

LAW OFFICE OF GEO. W. SMITH,
No. 86 WASHINGTON ST., CHICAGO, *Feb.* 21, 1871.

Mr. EDWARD POWERS :

Dear Sir—I have your letter of yesterday. In reply, I was present at Perryville, Stone River, Chicamauga, Mission Ridge, the various engagements between Chattanooga and Atlanta, Franklin and Nashville.

I remember that rain followed most of the above named battles, and particularly Perryville, Stone River and Nashville. Chicamauga was succeeded by a dense fog.
Yours, truly,
GEO. W. SMITH.

Doc. No. 9.

From General Wm. Vandever, of Iowa.

DUBUQUE, *Oct.* 15, 1870.

ED. POWERS, Chicago, Ill. :

Dear Sir — Your favor, of the 13th inst., with enclosed newspaper article received, and read with interest.

I do not think that I can give any particular information to guide you in your investigations. My observation, however, during the war, satisfies me that your theory is correct. Great battles were generally followed by storms of rain. This peculiarity was often the subject of comment in the army.

If, from such facts, you can suggest any feasible mode of bombarding the clouds and bringing down rain, the country will be indebted to you.

Very truly, yours,

WM. VANDEVER.

Doc. No. 10.

Extract from a Letter from the Adjutant General of the State of Ohio, dated Columbus, Oct. 17, 1870.

* * * * Your theory has always been a pet "hobby" of my own, and my observation, during the late war, led me to believe in its correctness. I have always noticed that heavy firing was followed by copious showers, with an uniformity which satisfied me that it was not mere coincidence. The best way to decide the matter, however, would be to institute a series of experiments during dry weather, when the barometrical signs indicated a continuance of drought.

Very respectfully, your obedient servant,

WM. A. KNAPP, *Adjutant General.*

Doc. No. 11.

Extract from a Letter from the Adjutant General of Wisconsin, dated at Madison, Oct. 26, 1870.

* * * I had occasion to notice myself that our battles were generally followed by rain during the war.

At Cedar Mountain, August 9, 1862, the battle was followed by a slight rain. The weather was, at the time, intensely hot, and the engagement was short, and but little artillery used. At the second battle of Bull Run, August 27 and 28, 1862, the rain poured in torrents. The closing engagement of that series of battles, that of Chantilly, was stopped by a drenching shower. At Chancellorsville, May 3, 1863, we had torrents of rain in about forty-eight hours after the cannonading was over. At Beverly Ford, June 9, 1863, we brought on rain by a sharp musketry and artillery fire, lasting half a day. At Gettysburg, where some three hundred cannon pounded from 12 till 4 o'clock, and musketry incessantly for three days, we had a night and day of pouring rain, setting in about six hours after the firing had ceased. Yours, respectfully,

E. E. BRYANT, *Adj't General.*

Doc. No. 12.

The foregoing letter fixes, approximately, the time of the commencement of the rain at the battle-field, after the battle of Gettysburg. The following extract from a letter from Colonel John Gibbon, 7th Infantry, U. S. A., speaks of the same rain as it occurred at a point some thirty miles southeastward.

"Immediately after the battle of Chancellorsville, there was a terrific rain storm, May 5, 1863. This was also the case after the battle of Gettysburg, the rain commencing to fall about twenty-four hours after the heavy cannonading of the 3rd of July; and at Westminster, about thirty miles from the battle-field, continued to rain heavily all night."

Doc. No. 13.

Mr. Abbott Mott, of the Engineer Department, U. S. A., in a communication to an officer, says of the commencement of the rain at Fredericksburg:

"At the battle of Fredericksburg, I was on the skirmish line the night of the retreat, and, consequently, was among the last to cross the Rappahannock on said retreat. I distinctly remember a very heavy rainstorm commenced while we were crossing on the pontoon bridge. This battle was notable for the amount and weight of ordnance used."

Doc. No 14.

From General J. A. Garfield, of Ohio.

HIRAM, O., *Oct.* 28, 1870.

EDW. POWERS, Chicago, Ill.:

Dear Sir — In answer to yours of the 22nd, I have to say that, while I did not take such observations as a scientific experiment requires, I did observe the frequent occurrence of heavy showers very soon after the battles in our late war. It was a matter much talked of in the army, and there was a general impression that the atmospheric disturbance, caused by heavy cannonading, hastened or created showers. I remember that heavy showers followed almost immediately after the battles of Shiloh, Stone River, Shelbyville, and Chicamauga. But, while these coincidences are curious and interesting, they are chiefly valuable from the fact that they challenge the attention of scientific men, and may lead to a discovery of causes which will prove valuable to our knowledge of meteorology.

Very truly, yours,

J. A. GARFIELD.

APPENDIX. 113

Doc. No. 15.

From Gen. J. M. Hedricks, of Iowa.

COURIER OFFICE, OTTUMWA, *Oct.* 28, 1870.

MR. EDWARD POWERS, Chicago, Ill.:

My opinion fully concurs with the theory of your article. I have, however, never taken time to investigate the phenomena sufficiently to give you an intelligent theory on the subject at present.

It is a highly interesting and important matter, and should be investigated.

In great haste, your ob't servant,
J. M. HEDRICKS.

Doc. No. 16.

From Gen. Jas. Barnett, of Ohio, late Chief of Artillery, Dept. of the Cumberland.

CLEVELAND, O., *Oct.* 28, 1870.

EDWARD POWERS, Esq.:

Dear Sir — I am in receipt of your favor of 22nd inst., enclosing your article from the *Evening Post*. It was a remarkable fact, which I think most of our army officers will recollect, that many of our battles were accompanied with rain, or rain immediately followed. Such was the fact at Pittsburg Landing and Stone River, and I think at other general engagements in our department. Chicamauga and Mission Ridge may have been exceptions, but of this I am not sure. Our advance from Murfreesboro, in which a good deal of artillery firing was done, was accompanied by rain all the way. I desire to talk with some of my army friends, who assemble here on the 24th ult., and will take pleasure in writing you further on the subject after we compare notes.

Yours,
J. BARNETT

The following letter from a distinguished citizen and late officer of volunteers will show what was the character of the weather in the Shenandoah Valley in the months of August and September, 1864; the document which follows it will show the apparent effect of the artillery skirmishing to which he refers, in tending to produce a change. A comparison of the two will also show that, in regard to the question of rain or no rain in that section after the battles of Winchester (Opequan Creek) and Fisher's Hill, fought September 19 and 22, it is unsafe, in speaking from memory, to affirm a negative. If there was rain, however, it was probably light, or the assertion that there was none would not be made with so much confidence. Rain followed both of these battles in the southeastern part of the same State, as before stated; though possibly there may have been no connection between the battles and the rains referred to.

Doc. No. 17.

Oct. 31, 1870.

Dear Sir — Your favor, with slip enclosed as to rain following battles, is at hand. There was a notion of the sort often mentioned in the army I belonged to. Off hand I have no opinion about it. Rain followed within twenty-four hours in the following cases:

  Carnifax Ferry, Sept. 10, 1861.
  Dublin Bridge, May 10, 1864.
  Winchester, July 24, 1864.
No rain after:
  South Mountain, Sept. 14, 1862.
  Lynchburg, June 20, 1864.
  Winchester, Sept. 19, 1864.
  Fisher's Hill, Sept. 22, 1864.
Other battles I dont recollect about.

On hearing of the attempt to investigate this subject by

APPENDIX. 115

you, the most conspicuous fact occurring to me was against the theory suggested.

In August and September, 1864, the Shenandoah Valley was the scene of unending warfare—daily battles—cannon firing from daylight to dark, and with it an unusual drouth. In September, 1870, no cannon-firing, and an unprecedented flood.

But my memory supplies too few facts to warrant an opinion.

Sincerely,
* * * * * *

Doc. No. 18.

Extracts from the Diary of Lt. W. Ashley, of Vaughn's Brigade, Co. C Battalion, Thomas' Legion, Wharton's Division, Breckenridge's Corps, Gen. Early's Army; who was killed at the Battle of Opequan Creek, near Winchester, Va., September 19, 1864.*

Newmarket, Va., Saturday, July 1, 1864. Daylight. Start through Edinburgh, * * * hot * * * .

July 2. Strasburg. Straggled and got a good dinner; encamped near Middletown.

July 3. Start through Newtown * * * .

July 4. Start to Martinsburg. Yanks had left in a hurry. Lots of plunder, * * * still hot and dusty.

July 5. Clear. * * * Marched to Potomac River, near Shepherdstown; waded it. * * *

July 6. Clear; still no rain. * * * Made foot of Maryland Heights about 11 P.M.

July 7. Cannonading all night; daylight start; we are now in position as reserve. Sharp fighting going on immediately in front; shells coming unpleasantly near every once in a while; passed over a man's foot in our road just now taken off by a cannon ball, suppose we are about one mile from their works. Harper's Ferry; dark; moved out over mountain to Rollersville by 2 A.M.; rain, rough and very dark.

* See Putnam's "Record of the Rebellion," Vol. XI. p. 153.

July 8. Clear. * * * Awful rain during the night; all and everything wet through. * * * *

(The diary shows no more rain until after the battle of Winchester, fought July 24, 1864.)

July 24. Clear; army in motion * * * * heavy shells and bullets coming thick among us * * * * drove the Yanks under a hot fire several miles through Winchester. * * *

July 25. Rain; all wet through and cold. * * *

August 17. Clear at daylight; ordered into front * * * ordered to charge the enemy; did so, under a heavy fire of artillery and small arms. * * * The fight was continued until 11 P.M. * * *

August 18. Rain. * * *

August 19. Hazy. * * * Skirmishing near Berryville. * * *

August 20. Rain. * * *

September 3. Cloudy. * * * Heavy artillery and musketry in direction of Berryville; rain; still fighting far away into the night.

September 4. Cloudy. Started to scene of last night's action * * * sharpshooters already engaged. 3 P.M. Flanked to left, and lay until night, endeavoring to draw them out to fight. They won't leave their intrenchments; bullets are whistling around us close. * * * Rain, cold and disagreeable.

September 5. Rain. Skirmishing heavy * * * heard firing in our front. * * * Rain falling heavy. * * *

September 6. Rain all day. * * *

September 9. Clear, cold night * * smart skirmishing. * * *

September 10. Rain. * * *

September 12. Rain. * * *

September 13. Clear; fighting on our left; * * fighting is winding to our right; * * * it is very heavy. 2 P.M. Cannonading heavy on our right.

September 14. Rain. * * *

September 15. Cloudy. * *

September 16. Rain. * * *

The facts in regard to the weather, set forth in the above document, must render the "conspicuous fact" noticed in the preceding document (I say it with all respect towards my correspondent) *less conspicuous* than, at first sight, it might appear.

Doc. No. 19.

From Major General S. P. Heintzelman, U. S. A.

NEW YORK, *Nov.* 6, 1870.

EDWARD POWERS, Esq., Civil Engineer, Chicago:

My Dear Sir — Your letter of the 6th of October, with its enclosure, I have received, and gladly contribute my mite towards the establishment of your theory.

I have been keeping a journal all my life, mostly a mere record of facts, and, as a general thing, I mention the weather. The enclosed notes have been carefully ex- extracted from this journal. I find that I have recorded, almost daily, the weather, and whether there was firing, from the first Bull Run, July 21, 1861, to September, 1862, when I was placed in command of the defences of the south side of the Potomac.

I have the impression that rain can be produced by the concussion of the atmosphere; and the only question in my mind has been, will it pay? It will depend upon the area of country that can be affected. Cannot some cheaper material be employed to produce the concussion than gunpowder?

A curious fact was brought to my mind the day after I received your letter. In conversation with a gentleman who moved on the Southern side, he inquired whether I had ever observed that, during the war in Virginia, there were no turkey buzzards in the vicinity of the armies. I recollect the fact, and attribute it to the great extent of atmosphere affected by the concussion of artillery firing, thus driving those timid birds away. This would go to show that the atmosphere is affected to a sufficient extent to make it practical, or that it will pay.

These observations have a greater value, as when they were recorded, I had no theory to sustain.

I am, sir, yours, truly,

S. P. HEINTZELMAN.

Doc. No. 20.

Notes from Journal kept by S. P. Heintzelman, commanding Third Army Corps, from July, 1861, *to September,* 1862.

July 21, 1861. This was a clear, hot day—the first battle of Bull Run. I reached my door, in Washington, the next morning, at 6½ A. M. " It commenced to rain a little before we got in."

Camp Winfield Scott, Yorktown, Va., Saturday, May 3, 1862. " Threatened rain this morning, but turned clear and pleasant." * * * " Some five hundred shots and shell were fired, yesterday, by the rebels. Not much firing to-day." * * " The rebels were very busy, till after midnight, firing " (artillery).

Sunday, May 4, 1862. * * " This is a beautiful morning." Rain commenced Sunday night. " It commenced raining in the night."

The battle of Williamsburg was fought Monday, May 5, 1862. It rained all day and into the night.* My impression is that it was clear the day after the battle.

Williamsburg, May 8 (Thursday), 1862. " A beautiful day." Wednesday was Franklin's affair at West Point.

Savage's Station, Saturday, May 31, 1862. " We had a very heavy thunder storm late in the afternoon yesterday,† and till in the night. It rained in torrents." * * It has been cloudy all day.

Savage's Station, June 1, 1862. " The clouds broke away early in the day, and it was warm."

Savage's Station, June 2, 1862. " Before daylight I got another dispatch from Marcy, to sustain Sumner with all my

---

* The rain that attended the battle of Williamsburg, was, probably, brought on by a sharp cannonade that took place in the afternoon of the day before. (See page 26) The reason why there was no rain the day after the battle is easily understood by supposing that the aqueous vapor within reach was, by that time, exhausted.

† This rain may have been caused by the battle of Hanover Court House, fought on the 27th, and the length of time that had elapsed between the battle and the rain, may have been owing to the fact that there had been a previous rain. See letter of Gen. Hagner, on page 130, in which notice is taken of the length of time which sometimes elapses between a *second* spell of firing and a second rain.

force. As I had already made arrangements for any contingency, I did not get up till it got light. It was then raining* a little. I dressed, and when the sun rose we had a rainbow. I think we will have a pleasant day."

Savage's Station, Tuesday, June 3, 1862. " Heavy rain and thunder storm last night. This morning has been hot. Mercury, at one time, in my tent, 94°." * * " All the wounded of my troops, and the prisoners, were sent off in the last train at 9 P. M. It commenced to rain pretty steadily before."

Savage's Station, Wednesday, June 4, 1862. " It rained more heavily last night and this morning, till about 9 A. M., than I have known for years. The whole country is flooded, both in the front and on the left. No enemy can move, even should he try, which I don't think he intends on this flank, after his defeat."

Savage's Station, Thursday, June 5, 1862. " It has been cloudy, and threatening rain with a few drops to-day "

Savage's Station, Friday, June 6, 1862. " It has been cloudy, and drizzled several times during the day. The weather is disagreeable enough."

Savage's Station, Saturday, June 7, 1862. " Cloudy this morning, but sun came out. In afternoon, a thunder storm, and now clearing off."

Savage's Station, Sunday, June 8, 1862. " This has been a pleasant day."

Savage's Station, Monday, June 9, 1862. " Quite cool, but pleasant, drying winds."

Savage's Station, June 10, 1862. " Rain† most of the night and this morning. Now a mist " * * " from top of a tree got a sketch. Two rebel flags were seen on a large building in Richmond, from this tree."

Savage's Station, Wednesday, June 11, 1862. " Cold last night. Rain ceased in the night, and pleasant to-day. High winds drying the roads rapidly."

* The rains of June 2, 3 and 4, followed the battle of Fair Oak or Seven Pines, fought May 31 and June 1, 1862. The heaviest of these rains, it will be noticed, fell on the night of June 3, and morning of June 4.

† Fremont's battles of Cross Keys and Port Republic, in Virginia, were fought June 8 and 9, 1862.

Savage's Station, Saturday, June 14, 1862. "A beautiful morning."

Savage's Station, Sunday, June 15. * * "We have some thunder and lightning, with rain, and the air cooled greatly From 93° to 66°. The enemy have been firing at our pickets, and we have lost some men, both in front of Hooker and Kearney."

Savage's Station, Monday, June 16, 1862. "Cool, but pleasant morning. Mercury, 57°. A great change since yesterday. We have had considerable skirmishing yesterday and to-day."

Savage's Station, Tuesday, June 17, 1862. "Cool night and all day. Mercury very little above 70°, and cool wind. * * The gunboats were firing near two hours to-day."

Savage's Station, Wednesday, June 18, 1862. "Cool last night and this morning, but getting warm again. * * About sundown there was some picket firing in front of Sumner, with rapid artillery firing. It lasted only a few minutes. * * Since dark, a heavy wind and rain."

Savage's Station, Thursday, June 19, 1862. "The rain, last evening, did not last long. This morning the roads are dusty. * * Cool morning, but warm day."

Savage's Station, June 21, 1862. "Warm and dusty to-day. It was unusually quiet all day, till late in the afternoon, when, suddenly, a brisk fire of musketry rang along Hooker's front, followed by artillery."

Savage's Station, June 22, 1862. * * "There was picket firing, at intervals, most of the night. At ten minutes before 3 A. M., several volleys were fired in rapid succession. * * but it only lasted a few minutes. * * We have had a few drops of rain since dark. * * Mercury has been 93° to-day, and little wind."

Savage's Station, Monday, June 23, 1862. "Quite warm till afternoon, and then showers of rain with a little thunder. * * All has been very quiet since yesterday morning. What can all this mean? * * It is quite cool since the rain, with some rain and lightning, and may rain more."

Savage's Station, Tuesday, June 24, 1862. * * "We had a very heavy rain storm, and thunder and wind, at midnight. The telegraph wires are down. * * At dawn, heavy musketry commenced, and soon followed by artillery. I thought it the attack, and had all up, but it did not last but

APPENDIX.   121

a few minutes. We afterward heard the rebels beat reveille. Had another heavy rain a little before night; has cooled the air much."

Savage's Station, Wednesday, June 25, 1862. "The rain made the morning and day pleasant."

[NOTE.—This is the affair of the "Orchards," in which my command lost some five hundred men, and pushed forward our pickets — the object of our attack.]

Savage's Station, Thursday, June 26, 1862. "For several hours this afternoon, heavy artillery firing has been going on, on our right. * * The firing is very steady and continuous, although it is getting dark. There must be quite a battle." [NOTE.—This was the battle of Mechanicsville.]

Savage's Station, Friday, June 27, 1862, 5½ P. M. "The battle still continues on the right." [NOTE.—This was the battle of "Gaines' Mills."]

Savage's Station, Saturday, June 28, 1862. "At 3 A. M., a heavy picket firing commenced, then joined in with artillery. * * There was, occasionally, artillery and some musketry firing during the forenoon. The enemy made a very determined attack on General Smith, and got into one of his redoubts. His infantry drove them out, etc., etc. * * I feared, this morning, it would rain." "In the night of the 28th, got an order to fall back to the lines I held May 31. * * * It was foggy." [NOTE.—On the 29th the battle of Savage's Station was fought.]

Junction Charles City and Quaker Roads, Monday, June 30, 1862. [NOTE.—Battle of Glendale fought.]

Thursday, July 2, 1862. * * "It commenced raining soon after light." [NOTE.—This was at Malvern Hill — the day after the battle.] * * "It was now about 6 A M., and raining hard."

Berkley's Farm, Thursday, July 2, 1862. "It rained hard in the night, and it is doubtful whether it will clear off now. * * At half past 10 A. M., the rebels commenced throwing shells into our camp, etc. * * It has not rained since morning, but it is not clear yet."

Near Berkley's Farm, Friday, July 4, 1862. "Clear sunshine. The roads and ground are drying rapidly.
*   *   *   *   *   *   *

Harrison's Bar, Tuesday, July 15, 1862. * * "Mercury was 96° to-day, but at dark a heavy thunder storm, and now down to 74°. * * There has oeen some gunboat firing down the river."

Harrison's Bar, Wednesday, July 16, 1862. " Mercury, 90°. In the evening, a heavy thunder shower."
*   *   *   *   *   *   *

Junction of Warrenton and Alexandria and Orange Railroads, Tuesday, Aug. 26, 1862. " The weather, the last few days, has not been very hot, but quite dusty."

Warrenton Junction, Wednesday, August 27, 1862. " There was some artillery firing in the night, and again this morning."

Manassas Junction, Thursday, August 28, 1862. " Some artillery firing on our left, at 8 A. M." [NOTE.—On the 27th was the affair at Bristow Station.] * * " We had quite a heavy shower as we passed Manassas Junction, but it only extended a short distance." [NOTE.—On the 27th, in the evening, about 9 P. M., we had a little rain. The ride from Warrenton Junction to Bristow Station, on the 27th, was very warm and dusty.]

Bull Run battlefield, near the Henry House, Friday, Aug. 29, 1862. " At 10 o'clock A. M., reached the field a mile beyond the stone bridge. Firing had commenced again." [NOTE.—This is the first day of the second Bull Run.]

Saturday, Aug. 30, 1862. [NOTE.—This is the second day of second Bull Run.]

Centreville, Va., Sunday, Aug. 31, 1862. At daylight it commenced raining. * * The rain did not last very long, but it is still cloudy. * * There was some firing this morning, but not much."

Fairfax C. H., Va., Tuesday, Sept. 2, 1862. On the day before, between Centreville and Fairfax C. H., " Heavy thunder and rain storm, at 6 P. M. * * After the rain, rode on a mile or so, and stopped opposite Kearney's left flank." [NOTE.—This was near Chantilly.]

Arlington, Md., Sept. 10, 1862. " There has been a heavy wind storm, but scarcely rain enough to lay the dust. Next day, rain showers all the forenoon."

S. P. HEINTZELMAN.

* The memoranda furnished concerning the time the army lay at Harrison's Bar, are, for the most part omitted, not being of special significance.

APPENDIX. 123

Doc. No. 21.
From Gen. John C. Starkweather, of Wisconsin.
SUNNY SIDE FARM, OCONOMONEE, *Nov.* 8, 1870.
EDWARD POWERS, Esq. :
Dear Sir — My house and its contents having just been destroyed by fire, prevents me (as to time) answering your favor in detail. I can therefore only say, in general terms, that I agree with you fully.
Yours truly,
JOHN C. STARKWEATHER.

Doc. No. 22.
From General Rob't A. McCoy, of Pennsylvania.
SURVEYOR GENERAL'S OFFICE, HARRISBURG, *Nov.* 14, 1870.
EDWARD POWERS, Esq., Civil Engineer, P.O. Box No. 45, Chicago, Ill. :
Dear Sir — Your favor relating to the subject of artillery fire producing rain, and requesting statement of my recollection as to rains following the principal battles of the late war, has been received. My time being very fully occupied by official duties, leaves me but little opportunity to make you a satisfactory reply, and the fact that I have not my memoranda book within consulting distance, compels me to write from memory.

The whole scope of my service in the army was in Eastern Virginia, Maryland, and Pennsylvania, with the army of the Potomac, and at no time exceeding 200 miles inland from the Chesapeake or Delaware bays.

This fact should be taken into consideration when considering the effects of our battles on the atmosphere or currents of air. My recollection is, that after the battle of Antietam — one in which much artillery was used — it rained the following day, the 18th September, 1862; that rain fell in considerable quantity on Tuesday after the Saturday of battle of Fredericksburg, Va., in December, 1862 — think some rain fell before Tuesday. A very large number of cannon were in position on the Stafford heights and the Fredericksburg and Mayre's heights; the valley between was densely filled with smoke from the discharge of cannon, small arms, and the burning of the town. The sight was grand. The sun appeared to toil through the density of smoke.

It rained on the third day, I think, of the Chancellorsville battle, May 3, 1863.

Very heavy rain fell after the battle of Gettysburg; in fact the night of the last day of the battle your correspondent was soaked with rain whilst examining and arranging outposts.

My recollection does not serve me as to the three days of the Wilderness battle, but in the execution of my duties I got very wet one day near the close of the battle of Spottsylvania C. H., 1864.

I omitted to mention that it rained very hard immediately after the second battle of Bull Run, August 31, 1862.

I am not prepared to go into any scientific argument for or against the theory, but believe that the concussions—jarring of the atmosphere by the sound, as well as disturbance of it by the smoke of battle, produced rain.

I am inclined to the opinion, that the smoke is not without its effect in producing rain, for I remember to have observed when a boy, living in the interior of this State, that in the early fall, the farmers who had cleared new lands usually burned the brush and log heaps about one time, causing dense smoke through the valley, and remember that rain usually followed; also after the burning over of mountain lands.

But, as I have before remarked, that all my experience and observation have been confined to a narrow limit not far from the sea, where greater moisture exists in the atmosphere than further inland, perhaps that fact might have had much to do with the frequent rains apparently caused by firing and by smoke. If concussion, or jarring the air, is the moving power, then the firing should be directly up in the air, or rather by batteries placed at say one or two miles apart, and fired into the air at a proper angle and towards each other.

Begging your pardon for inflicting upon you so crude a letter, I have the honor to be,

Very truly, yours,

Rob't A McCoy,
Late Ass't Adj't Gen'l 3rd Div. 5th Army Corps, Army of Potomac.

Doc. No. 23.

From General J. M. Campbell, of Pennsylvania.

JOHNSTOWN, Pa., *Nov.* 16, 1870.

ED. POWERS, Esq.:

Dear Sir— Your favor, of the 7th inst., with enclosed slip, I find awaiting me on my return home.

At present I can remember but two battles during our late war, which were closely followed by rain. The first, after the battle of New Market, Va., on the 15th of May, 1864, the other after the battle of Winchester (Crooks, Va., July 24, 1864. There were, doubtless, others, but I cannot recall them with distinctness. I have heard the idea you advance frequently discussed since the war, and believe there is "something in it."

Very respectfully, yours,

J. M. CAMPBELL.

Doc. No. 24.

From General E. L. Dana, of Pennsylvania.

WILKESBARRE, *Nov.* 21, 1870.

EDWARD POWERS, Esq., C. E.:

My Dear Sir — Since the receipt of your favor of the 7th inst., I have been engaged in holding court, with no leisure, until this evening, for a reply.

I showed your article to several military gentlemen of this town, who concur, in their recollection, that nearly every great battle, of the late war, was either attended, before its close, or immediately followed, by a heavy fall of rain; generally with much electric action. The occurrence was the subject of remark. I think on the third day after the commencement of the Chancellorsville movement, and in the midst of a rapid cannonade, there came on a fearful thunderstorm, and, for a time, we were at a loss, in the thick woods, to distinguish the "artillery of heaven" from that of earth.

At Gettysburg, on the 4th of July, the day following the decisive conflict of the 3rd, characterized by the heaviest cannonade of the war, there was a severe storm, a large quantity of water falling. There was a slight fall of rain during the battle of the 1st of July, at Gettysburg, and in the evening.

In one of the operations before Petersburg, I think in October, 1864, which was there called the Squirrel Level Road, a heavy rain followed immediately after the action. The same coincidence occurred in the two actions at Hatcher's Run.

The rain, which fell at Chancellorsville, August, 1863, to which I have referred, you may recollect, was such as to raise the river and threaten our pontoon bridges, and, probably, hastened our re-crossing the river.

These meagre reminiscences touching the question, suggested in your letter, with those of General Osborne, to whom I showed your article, are all that I am able to recall distinctly at present. We had many rainstorms, of course, at other times.

The mountains around our valley of Wyoming occasionally take fire, and, after a day or two of burning, form nearly a circle. A rain opportunely occurs, about this time, and extinguishes it. Prof. Espy, some years ago, had a theory of the effect of fires in producing rain.

I am, very truly, etc.,

E. L. DANA.

With the foregoing letter, General Dana was kind enough to forward the following from General E. S. Osborne:

Doc. No. 25.

WILKESBARRE, *Nov.* 19, 1870.

Gen. E. L. DANA:

General — The letter of Mr. Powers to you, and also the article containing remarks relative to the supposed effect of artillery fire in producing rain have been read by me, and in compliance with your desire, I would state that heavy storms followed the following battles, viz.: Chancellorsville, Wilderness, North Anna River, Weldon Railroad, and Hatcher's Run. Upon these occasions, I am positive that we had very heavy rain, accompanied with thunder. Whether other bat-

tles, in which the Army of the Potomac was engaged, were followed by storms, I do not now distinctly remember.
Very respectfully, your obedient servant and friend,
E. S. OSBORNE.

Doc. No. 26.

From Brevet Major General Henry J. Hunt, U. S. A.

FORT ADAMS, NEWPORT, R. I., *Nov.* 13, 1870.

Mr. EDWARD POWERS, Chicago, Ill. :

Dear Sir — Your note of the 18th of October, with its enclosure, reached me in due course. My absence for a portion of the time since, and other duties, have prevented my returning an earlier answer.

I cannot, at this time, give very accurate answers to your questions respecting the occurrence of storms after battles, but, in many cases, I can remember, with sufficient certainty, their occurrence, and very nearly the period within which they occurred.

The battle of Cherubusco, in the valley of Mexico, was fought on the 20th of August, 1847. The rainy season must have been closed, or near its close. At Puebla, during the months of June, July, and, perhaps, the beginning of August, there were heavy falls of rain *every afternoon*, the skies clearing before sunset, and the atmosphere being remarkably clear until the *next* afternoon ; but I remember that on the march from Puebla, which commenced 7th August, the days were, generally, if not always, clear, bright and beautiful. On the 14th, the whole day was bright and clear. I was specially engaged that day on duty, which I remember. I believe the 15th was a similarly clear day, as was the 16th, the date of the commencement of the movement round Chalco. There was some little rain on the 17th or 18th, but it was not, I believe, very heavy. The 19th was clear and beautiful, in the afternoon, at the usual hour, for rains. I remember that I was watching the movement taking place at Contreras, from a distance, at the time. I think that night was cloudy, dark, and, perhaps, rainy. I do not think the rain was very heavy. There had been cannonading at Contreras during the day.

Since writing the foregoing, I have found General Scott's

report of this action. He describes the fire of artillery as *heavy*, the enemy having twenty-two guns mounted, to which we could only reply with a battery or two of six pounders, and one of mountain howitzers. I doubt if all the enemy's guns were used, or could be brought to bear that afternoon. However, General Scott says, in one place: " It was already dark, and the cold rain *had begun to fall in torrents* on our unsheltered troops." He afterwards refers to the night march of the troops being hindered by "mud and rain." I was under partial shelter that night, which may account for my recollections not being very clear of "torrents of rain."

The battle of Churubusco was fought the next day, which was bright and clear; I don't remember rain. The day after it rained heavily whilst we were on the march to Tacubaya. I do not remember, with sufficient distinctness, the condition of the weather after other Mexican battles; nor, considering the nearness of the period named to the rainy season, are the above facts, perhaps, of great significance.

The 21st of July, 1861, the day of the battle of Bull Run, was clear, hot and bright all day long. The next afternoon there were "torrents of rain," which continued all night.

The battle of Gaines' Mill was fought June 27, 1862. It was a bright, clear day, as was, also, the 28th; but, on the night of the 28th, and morning of the 29th, it rained heavily.

The 29th and 30th of June were fair, bright days. The battle of Malvern was fought July 1, a bright, clear day. During the night it commenced raining; and on the 2nd, and 3rd, also, I think, it poured down.

I do not remember that the battle of Antietam was followed by rain. It may have been; I do not remember. Nor can I speak, positively, as to that of Fredericksburg, December 11–13, 1862.

The battles of Chancellorsville, May 2–4, were fought, I believe, thoughout in fair weather. In the afternoon of the day the army recrossed (5th), it poured rain — raising the river, sweeping off the bridges, so cutting off the movement until they could be restored — and continued all night, and part of the next day.

The battle of Gettysburg was fought July 1, 2 and 3, in clear weather. On the 4th it rained furiously, and continued part of the 5th.

I do not remember that violent rain followed the battle of the Wilderness. There was not much artillery fire compared with the magnitude of the forces engaged. This was on the 5th and 6th. There was rain, I believe, on the 8th or 9th, during the first fight, on the right, at Spottsylvania C. H. After the heavy fighting there, the army, on its move to the left, to renew the attack, did so through a heavy storm of rain, which continued next day.

The fighting, however, from the 4th of May to the 27th, when the army crossed the Pamunkey, was so continuous, that little, if any, conclusion, from the rains that happened in that period, could be drawn, as affecting the question of *cause.*

As to the temperature and direction of the wind, at these times, I cannot give you any information worth recording. The occurrence of rains, soon after battles, I have noticed frequently; but whether previous statements that such was often the case, or whether the frequency of the occurrence attracted my attention, I cannot say. Frequently without means of keeping memoranda on pressing subjects, of course I had neither time nor opportunity to record such facts. Indeed, the necessity of trusting to memory for many things is what enables me to recall circumstances of time, place and weather; that permit me to write this letter, which, I fear, you will not find very useful, but it is the best I can do to comply with your request. I regret exceedingly that I cannot do better.

Very respectfully, your obedient servant,
HENRY J. HUNT, *Maj. Gen. Bvt.*
Late Chief of Artillery, Army of the Potomac.

Doc. No. 27.

From Brevet Brigadier General P. V. Hagner, of the Ordnance Department, U. S. A.

WATERVLIET ARSENAL, WEST TROY, N. Y., *Dec.* 28, 1870.

EDWARD POWERS, Esq., Box 45, Chicago:

Sir — I have received your note of the 24th, and respond with pleasure to your inquiries. I have no doubt that heavy firing of artillery is, almost invariably, soon succeeded by a fall of rain; and I think it will be proven that this effect is due to some other cause than the heat evolved in burning gunpowder. It would seem, also, pretty certain that a *second* spell of firing, in the same vicinity, will not produce a *second rainstorm* within a day or two (or more) after the first. You will have a good chance of deciding the exact amount of influence due to this by observing carefully the reports from the Prussio-French battle-fields. The matter is alluded to under the head of " Rain following the discharge of ordnance," in the "Annual of Scientific Discovery," page 392, year 1862, and page 333, year 1863.

All accounts of the battle of Waterloo tell of the heavy rains during that battle. The same is true of many others of Napoleon's battles.

About the battles of Mexico, concerning which you ask my remembrance, I can refer you to Henry B. Dawson's " Battles of the United States," where, on page 467, siege of Monterey, September 21, 1846, after a continuous engagement, " Soon after the storming of the two forts, Federacion and Soldado, a violent storm came up;" and page 468, " the men were exposed to the unbroken pelting of a pitiless storm during the night." Also, page 473 : " General Worth and second division, as has been seen, spent the night entirely exposed to the peltings of a severe storm."

Battle of Buena Vista, page 491, the firing commenced on the morning of the 22nd of February, and "at night the cold wind and drizzling rain which chilled the bodies." But there was heavier firing on the 23rd, while, at night, the " moon shone ;" page 497.

Battle of Contreras, page 563 : " The battle raged furiously, and for more than three hours the entire force was under fire." * * * " Night, at length, put an end to the conflict, and a cold rain, which soon afterwards began to fall in torrents,"—(as I well remember.)

I am almost certain that in the afternoon and evening of the 8th of September, after the battle of Molino del Rey, there was a hard rain. It was clear until 1 or 2 o'clock, I remember.

We fired all day of the 12th of September, at Chepultepec, but not very rapidly (as we could not spare many shot.) It was clear the 13th, but, I think, rained before night, on the 12th. (I do not feel certain, however, and cannot now confirm my impression.) It was dark and cloudy the night of the 13th, (when I was throwing some shells and shot from San Cosme Garita to let the Mexicans *feel where we were,*) but bright enough the morning of the 14th, when we marched into the city.

I am sorry that I cannot be more definite in my information.

Very respectfully, sir, your obedient servant,
P. V. HAGNER.

General Hagner, in a subsequent letter, mentions that he thinks he has a distinct recollection of rain, which occurred after the battle of Chepultepec, between the time he was firing from the Garita of San Cosme, and sunrise, on the morning of the 14th.

Doc. No. 28.

From Major General Thos. J. Wood, U. S. A.

DAYTON, O., *Jan.* 9, 1871.

EDWARD POWERS, Esq., Civil Engineer, Chicago, Ill.:

Dear Sir — Your note of the 28th ult., covering a slip from the Chicago *Post*, written by you, is received.

The theory, suggested by you, of the relation of cause and effect between great atmospheric disturbances, such as are caused by the heavy cannonading in great battles, and the occurrence of rain immediately afterwards, is not new; but the suggestion of a series of experiments, with a view to the determination, with reasonable satisfaction, whether the theory is true, for the purpose of making it practically useful, is novel, and well worthy of consideration.

A collation of facts, drawn from many reliable sources, might well serve as the basis of further experiments

Many battles, as all know who have had any experience on the subject, have been followed by rain, while others have not. This fact would seem to indicate that if the atmospheric disturbances caused by the firing in battle have any effect in producing rain, the actual accomplishment of rain depends, in a great measure, if not chiefly, on the condition of the atmosphere. The condition of the atmosphere should, hence, be one of the chief factors to be observed in the experiments you propose.

With these preliminary remarks, I will give you a few facts, drawn from my own personal experience.

Battle of Monterey, September 23, 1846. Morning bright and fair, with no indications of rain. Heavy cannonading during the day. The evening and night closed in with heavy rain.

Battle of Contreras, August 19, 1847. Same remarks applicable as to battle of Monterey.

Battle of Shiloh, April 6, 1862. Same remarks as to Monterey.

Battle of Stone River, December 31, 1862. Much heavy cannonading, followed by sleet, snow and rain.

Battle of Nashville, December 15 and 16, 1864. Same remarks applicable as to battle of Stone River.

I might mention similar facts, drawn from my own experience or historical reading, but these, with such as you will, doubtlessly, derive from like sources, will, probably, answer your purpose.

Very respectfully, etc.,
TH. J. WOOD,
Major General, U. S. Army.

Doc. No. 29.

From Major General R. W. Johnson, U. S. A.

SAINT PAUL, MINN., *Jan.* 10, 1871.

My Dear Sir—Yours. of the 28th ult., with enclosure, is received. Throughout the late war I had frequent occasion to observe that heavy cannonading was soon followed by rain. I was present and engaged in the battles of Stone river. Liberty Gap, Chicamauga, Mission Ridge; the campaign. to within three miles, of Atlanta; and, also, the battle of Nashville. Heavy rains followed Stone River. Liberty Gap, Mission Ridge, and Nashville. During the Atlanta campaign, which was a continuous battle of ninety days, we had heavy rains, at short intervals. After the battle of Chicamauga no rain fell; but it must be remembered that this battle was fought in the woods, where artillery could not be handled easily, and there was but little cannonading on that field. It was so common for rain to succeed battles, that I think it was generally conceded that these showers were brought about by the heavy firing.

In my own opinion, I am satisfied that rain can be produced by a heavy cannonading. My own experience satisfies me, and I think the opinion became general during the war.

Your obedient servant,
R. W. JOHNSON,
Major General, U. S. A., retired.

Doc. No. 30.

From Major General Schuyler Hamilton, of New York.

NEW YORK CITY, *Jan.* 14, 1871.

EDW. POWERS, Esq., P. O. Box No. 45, Chicago, Ill.:

Dear Sir — Your favor of December 24, 1870, was only received yesterday. You will see, by the enclosed envelope, why. As to the subject of rain after heavy firing in battle, I can say, as to Monterey, that, though the day on which the battle commenced was, in the morning, bright and beautiful, a heavy rain fell in the evening, viz., September 21, 1846. I think the same phenomena was exhibited September 22 and 23. I was so grievously wounded at the time of the battles of Molino del Rey and Chepultepec, as to be unable to participate. However, at Mira Flores, the affair in which I was wounded, where the firing of small arms was very brisk for a time, a bright afternoon and day was followed by a heavy fall of rain. I have referred your note to Col. H. L. Scott. who was Chief-of-Staff and Adjutant General to Gen. Scott in Mexico, asking him as to his recollections as to Molino del Rey and Chepultepec. I merely state my recollection as to the fact that rain fell on the occasions referred to by me. I think my observation has been wherever I have been engaged, that the concussion produced by the heavy fire of a battle has been invariably followed by rain. Such was the case after Palo Alto, Mex., also.

I remain your obedient servant,

SCHUYLER HAMILTON.

In referring the writer's inquiries to Colonel Scott, as General Hamilton was kind enough to do, the following correspondence ensued:

Col. H. L. Scott will oblige me by stating if he has any recollection about the weather after Molino del Rey and Chepultepec, Mex., as I wish to oblige the writer of enclosed by a simple statement of the fact of rain or no rain — leaving to him his theory.

Yours, truly,

SCHUYLER HAMILTON,
Late Major General of Volunteers.

DEAR HAMILTON — I am unable to recollect whether it rained or not after Molino del Rey, and I probably should not be able to recollect how it was after Chepultepec and the City of Mexico, but in the "Mexican History of the War in Mexico," I find the following passage: "The morning of the 14th (September) was as gloomy and sad as the destiny of the Republic. There was a mist so thick that objects could not be seen at a few steps distance. Soon after a light shower began to fall, which soaked the soldiers, and the cold increased that was felt."

Truly, yours,

H. L. SCOTT.

Doc. No. 31.

From Major General John C. Robinson, U. S. A.

BINGHAMTON, N. Y., *Jan.* 16, 1871.

EDW. POWERS, Esq., Chicago:

Sir — Your favor, of the 28th ult., forwarded from Washington, has been received. In reply, I would say that I have not the slightest doubt of the correctness of the theory you mention. I have observed that all great battles in which

I have been engaged, (particularly those of several days' continuance,) were followed by heavy rains. Some of the battles in Mexico, the battles on the Chickahominy, the seven days' battles, the battle of Fredericksburg, and the battle of Gettysburg were immediately followed by very heavy rains. That heavy cannonading will produce rain, does not, in my opinion, admit of doubt.

Very respectfully, yours, etc.,
JOHN C. ROBINSON,
Major General, U. S. A.

Doc. No. 32.

From Major General J. M. Schofield, U. S. A.

SAN FRANCISCO, CALIFORNIA, *Jan.* 19, 1871.

Mr. EDWARD POWERS, Chicago, Ill. :

Dear Sir — Your letter of October 15, has, unintentionally, been left unanswered until now. I cannot attempt to give, from memory, specific facts which would be of value to you, but the general fact of a fall of rain during or immediately following heavy discharges of artillery and musketry, has been, in my experience, so common, and regarded so much a matter of course, as to attract no special notice in individual cases. My impression has been, however, that this phenomenon results only when the quantity of moisture in the atmosphere approaches nearly the point of saturation, and when any considerable disturbance of equilibrium might, naturally, be expected to produce condensation. In a calm, moist atmosphere, heavy discharges of artillery are, I think, very generally followed immediately by a fall of rain. Beyond this my experience does not enable me to express an opinion.

The subject you have under consideration is one of much interest, and may prove to be of no little importance.

Very respectfully,
J. M. SCHOFIELD.

Doc. No. 33.

Extracts from a Letter from Major H. S. Melcher, dated Portland, Maine, Feb. 18, 1871.

"Antietam," September 17th, 1862, was the first battle I was in. The first day's fighting was sharp, with heavy artillery firing; the next day there was a very sudden and heavy shower of rain; had been none for five days previous."

"Alder," June 21, 1863. A general skirmish, with but little artillery firing. Next day foggy, with quite a fall of rain. Had been very dry for two weeks."

Major Melcher also mentions the rains after Fredericksburg, Chancellorsville, and Gettysburg; but as these are elsewhere sufficiently described, his description is omitted. In regard to the battle of Spottsylvania, he says: "A very heavy rain storm set in the night of the 10th."

Of subsequent battles:

"Being wounded, I did not rejoin the army till October, so that I cannot speak of the results of operations in front of Richmond and Petersburg; but at the battle of Dabney's Mills, February 6, 1865, where considerable artillery was used, afternoon of the 6th, a storm of rain and snow set in next morning."

"The first day's operations in front of Petersburg, which resulted in the fall of that place and final overthrow of the rebellion, was followed by a heavy rain storm, which continued all night and the day following."

Doc. No. 34.

From Colonel R. Kennicott, of Illinois.

CHICAGO, ILL., *Feb.* 22, 1871.

Dear Sir—Yours, of yesterday, is at hand. In reply, I have the honor to state that I was at the battle of Pea

Ridge, Ark. It did rain after that battle; I think the morning after, March 9, 1862, when a very heavy shower fell.

I did not march with the command up Red River, and did not belong to the army or armies engaged at the other places you mention, save at Vicksburg. I was present there from June 14 to the fall, but do not remember about the rainfalls, though I think there were several light showers. I think we had rain just after Prairie Grove, and I have several times noticed that cannonading was followed by rain.

With regrets that I have no data with which to furnish you,

I am, sir, very respectfully, yours,

R. KENNICOTT.

Doc. No. 35.

From Rear Admiral L. M. Goldsborough, U. S. Navy.

NAVY YARD, WASHINGTON, D. C., *Feb.* 25, 1871.

EDW. POWERS, Esq. :

Dear Sir — In reply to yours of the 22nd, received by the mail of yesterday, I have to say that my impression is quite decided upon the subject to which you advert, but it is impossible for me, at this time to furnish you with the details you wish without a research, which I have not now the time to make. It is my firm belief that, invariably, an early fall of rain follows a heavy firing of artillery, continued for a few hours in a limited district of space. It may be, however, that the phenomenon is more likely to occur on land than at sea; and I am inclined to think that such is the case.

As well as I can now recollect, rain occurred the next day after the bombardment of Roanoke Island, if not during the night of the same day, February, 1862. But to get the facts you want, with precision, I would commend you to consult our Log Books. They are carefully kept preserved in our Bureau of Navigation, and they give the weather, at short intervals, for every day of the year, recorded, too, in the most systematic manner. In a word, they will tell you, beyond all doubt or dispute, exactly what weather did occur after every naval engagement. * * * They are a source to which you should appeal for the most reliable information in regard to the subject you have in hand, which, to my apprehension, is fraught with interest, and can be worked up, probably, to the advantage of science, if not to special benefit. Rain, for instance, as we all know, is the best of fertilizers; and a means within general reach may be discovered to cause it to descend when most wanted. Philosophy holds all things to be possible.

Very truly, yours,
L. M. GOLDSBOROUGH,
Rear Admiral, U. S. Navy.

Doc. No. 36.

From General Julius White, of Illinois.

CHICAGO, ILL., *Feb.* 26, 1871.

EDWARD POWERS, Esq., C. E., Amenia, N. Y.:

Dear Sir — In reply to your note of the 20th inst., asking the result of my observations, during the late war, relative to the theory that rains are produced by the firing of artillery, I would state that the only marked instance within my recollection occurred in the month of August, 1864, at about the time the Weldon railroad was taken by the Fifth Corps, under General Warren.

During the fighting which ensued upon that event, say from the 18th to the 26th, within which there were two battles fought by the Fifth, and a part of the Ninth Corps, and one about five miles south by the Second Corps, I noticed and called the attention of some of the officers with whom I was associated, to the fact that the sun rose and set for a number of days upon skies which were free from clouds, yet the rain fell copiously during the nights.

It was regarded as remarkable, if not anomalous, and the theory to which you allude was somewhat discussed, at the time, in connection with the fact.

It is proper to state that there was one day (the 21st), when a heavy fog prevailed — brought to my recollection by the fact that the enemy attacked on that morning.

The effect upon the health of the troops, and especially

upon my own, gave me further reason to remark the state of the weather, and I attribute a subsequent long sickness to the extreme heat, during the days, and copious rains of the nights, during the period mentioned.

Very respectfully, yours,
JULIUS WHITE.

Doc. No. 37.

Extract from a Letter from Commander E. Barrett, U. S. Navy, dated Ordnance Office, Navy Yard, New York, March 1, 1871.

" From boyhood I noticed that the atmosphere was affected by the firing of heavy ordnance. My attention was first attracted to the subject in 1843 and 1844, in the harbor of Rio de Janeiro. We had had beautiful weather: a change was brought about by the arrival of the Princess of Naples, now Empress of Brazil. She was accompanied by the Neapolitan and Brazilian squadrons. On her arrival the fortifications and foreign squadrons began to fire. The firing continued for an hour or more, when the sky was suddenly obscured, and heavy showers followed. The next day was calm and partly overcast; as soon as the firing of salutes was renewed, the rain began to fall, and the breeze sprang up."

As I have already given, or quoted from, some letters expressing a different opinion from the majority of the foregoing, I trust it will not be supposed to be my intention to convey the idea that the opinion became general in our army, during our late war, that artillery firing brought rain. Such, indeed, was not the case, as there were a vast number who gave no thought to the subject; and there are some who doubt that this effect was ever produced. In order to show fairly what is said on both sides of the question, the following additional opinions, adverse to the theory under discussion, are here introduced.

General J. H. Wilson, U. S. Engineers, a gentleman of high scientific attainments, as well as of fame as a soldier, says: "I am constrained to say that my experience, extending from the first to the last days of our late war, during which I participated in nearly all of our great battles, does not justify me in pronouncing an opinion favorable to your hypothesis in reference to the influence of cannon firing in producing rain. I should add, however, that I have given but little attention to the subject, and, therefore, do not wish to be understood as saying that you are incorrect in your suppositions. The question, although not a new one, is, as you justly remark, one of great interest, and should be settled by experiments directed solely to that end. I do not regard the casual recollections of officers in reference to such a matter as of any great value. A well directed series of experiments would be of infinitely more service towards the formation of true opinions. Trusting that your investigations may be so encouraged as to enable you to arrive at the truth, whatever it may be, I am," etc.

Another officer, who is more decided in his disbelief that battles cause rain, says: "Most of our battles were fought in the season of thunder storms, and were, almost without exception, preceded by extremely hot, sultry weather—so were the battles of Chancellorsville and Gettysburg, and many of the engagements during the Atlanta campaign, by an atmosphere which made it almost impossible for the troops to reach the battle fields, and I considered the then following severe thunder and rain storms as the natural course of nature. The Atlanta campaign was especially remarkable for such storms, which, almost twice a week, took place; but I remember distinctly that they not always followed heavy cannonades or musketry. In some instances, as towards the end of the series of battles for the possession of Kenesaw Mountain, the weather changed to the fairer, to remain so for almost two weeks, in spite of tremendous firing, while rain had been impending, and fell afterward on the march to that position, and during the commencement of the engagement. The second battle of Bull Run was fought in fair, hot weather, which remained until our retreat into the defences of Washington, some days after. So was Pope's campaign only interrupted by one severe thunder storm.

"The campaign of Fremont through the Shenandoah Valley was preceded and accompanied by thunder storms. Missionary Ridge was fought on a beautiful day, and the clouds which covered Lookout Mountain were evidently dispersed by the heavy cannonades, a similar instance to which I observed afterwards at Charleston, East Tennessee.

"My general observation is, that in 1862, fair weather; and in 1863, '64 and '65, rains prevailed, during and after battles. In all cases, however, I feel rather inclined to believe that these rains were the consequence of the already existing condition of the atmosphere, and the season of the year."

In answer to the statements contained in the letter last quoted from, I refer, first, to the foregoing pages for proof that not only in the season of thunder storms, but in all seasons, rain, snow, or hail follows battles. In relation to the Atlanta campaign, which has been described as a continuous battle for ninety days, I remark that the admission that it was remarkable for thunder storms is good evidence in support of the theory under discussion, and no less so that it is made to appear that a little time sometimes elapsed between a spell of heavy firing and the succeeding rain, or that rain sometimes followed the light rather than the heavy firing.

In regard to the battle of Mission Ridge, admitting that the clouds were dispersed, it is safe to say that they very soon after gathered again, for this battle is said to have been followed by rain. Admitting, also, that Fremont's campaign in the Shenandoah Valley was preceded by rain, it is equally true that it was preceded by a battle—viz., General Banks' battle of Winchester. The facts which I have given, with the evidence by which they are supported, show that my correspondent is in error in supposing that the occurrence of rain, in connection with battles, was less marked in 1862 than in 1863, '64 and '65. That his memory is at fault in some other points also, is shown by the fact that rain immediately followed each and every engagement of General Pope's Virginia campaign, and by the further fact that, so far from there having been no rain after the second battle of Bull Run until the retreat of our army into the defences of Washington, there was a shower the next morning following the close of the battle, and on the second day after it, during the battle of Chantilly, a most terrific thunder storm.

The above, and the other extracts which have been given from letters adverse to the theory under consideration, are selected, not for the ease with which they can be answered, but because of the high sources from which they come, and because they represent the principal points that have been made against it as far as my correspondence has extended, except one, which will be presently noticed. The few who express doubt as to the fact of the production of artificial weather by battles or artillery firing, generally base their opinions upon what can be shown to be unsound premises; and, indeed, it is not surprising that this should be the case. Those who gave little or no attention to the subject during the war, when matters of more instant importance generally, after battles, required their care, must naturally be more or less at fault in their recollections of the weather at such times, after so long a period as has since elapsed.

Col. C. H. Crane, Assistant Surgeon General, U. S. A., says: "If it could be shown that rains were decidedly more frequent immediately after battles than antecedent probability would lead us to expect—that is, for instance, if the day after a great battle was rainy in sixty cases out of an hundred, while the average probability of a rainy day, in the places where the battles were fought, at that season of the year when they were fought, were only twenty per cent.,—it would then remain to be inquired into whether battles were not commonly preceded by a number of days of dry weather that made military movements more active and brought the armies together."

In answer to the objection, which, I understand, to be here suggested, it is not necessary to dwell upon the fact that, during our late war, orders for military movements, at distant points, frequently emanated from Washington, where the state of the weather, at the time existing at such points, could not be properly considered; for, if it be true that battles are generally preceded by several days of dry weather, this circumstance would be rightly regarded by most persons as furnishing an argument *for* the theory that maintains that the battles cause the succeeding rains, rather than one against it. At any given time during a spell of dry, settled weather, it is more reasonable to expect that the next day will be fair, than that it will be rainy. If rains naturally occurred at regular intervals, then, in considering whether a rain following a battle was or was not produced by the battle, it would be necessary to inquire whether it was not the time for rain, though there had been no artificial cause to produce it—but, occurring at irregular intervals, as they do, this point would seem to be one which has no material bearing on the question. This will be more apparent, if we consider how it would affect the credibility of the theory, if it could be shown that battles were generally preceded by wet weather, instead of dry. Indeed, in the case of two or three battles, it has been mentioned as a fact bearing against it, that they were preceded, as well as followed, by rain. Thus it is seen that while one individual would doubt that a battle caused the rain which followed it, because there had been previous dry weather, and it was time to expect rain, another would entertain the same doubt in reference to another battle, because there had been rain immediately previous, and it was reasonable to expect more. Such reasoning requires only to be stated to show its fallacy. From the mere fact of dry or wet weather before a battle, it cannot be predicated what should be the weather which follows it.

I have already presented evidence sufficient to establish the facts that have been stated, and in the following letters will be found still further testimony bearing on some of the same points.

Doc. No. 38.
From Captain N. J. Manning, 23rd Ohio Vol. Infantry.

BARNESVIILE, O., *Oct.* 31, 1870.
EDWARD POWERS, Esq., Civil Engineer, Chicago, Ill.:

Sir—I noticed an article in the New York *Evening Post*, entitled "Artillery firing and rain," signed by you, and requesting the experience of any one who had observed the same; and I, in response to that, will give you some of my experience and observations.

I was a member of the 25th Ohio Vol. Infantry from the 10th of June, 1861, until the 27th of July, 1864, and participated in all the engagements the regiment was in between said dates (excepting the taking of Fort Wagner, on Morris Island, in front of Charleston, S. C.,) to wit: Cheat Mountain, Green Brier, Allegheny Summit, Monterey, Bull Pasture Mountain, in West Virginia; the pursuit of Jackson, by Fremont, up the Shenandoah Valley, which ended in the battle of Cross Keys; Cedar Mountain; Pope's retreat, which culminated in the second battle of Bull Run; Fredericksburgh and Chancellorsville, in Virginia; and last, not least, the battle of Gettysburg, in Pennsylvania — in all of which engagements, or wherever there was artillery practice of any moment, I observed that rain fell either during the engagement or immediately thereafter, and the quantity of rain seemed to be in proportion to the amount of artillery firing, and I thought then, and I believe now, that the firing caused the rain.

The rain falling, on all the foregoing instances, without a single exception, convinced me that it could not be merely a coincidence, but that the rain was brought on by the firing, and I think there is no doubt of it. I also heartily concur with you in your views, that, in time of drouths, large amounts of money could be made to the country, at a little expense, by the use of powder in that manner.

Yours, respectfully,
N. J. MANNING,
Late Captain 25th O. V. I.

A similarity will be noticed in the ideas of the writer of the following with those of Captain Manning, in relation to the amount of rain as compared to the amount of firing.

Doc. No. 39.

NEW YORK, *Oct.* 17, 1870.

Sir — I notice your article in to-day's *Telegram*, and believe its theory is correct. Have thought so since 1861, and my idea was confirmed by every heavy cannonading or musketry fire in my vicinity.

There were quite heavy rains very soon after the fight at Big Bethel, the naval battle in Hampton Roads between the Monitor and Merrimac, etc., etc., and the very severe battle at Malvern Hills.

It struck me as a curious fact that the amount of rain which fell after each battle, seemed to be very nearly in proportion to the amount of powder that was burnt.

Respectfully,
FRED. M. PATRICK,
Of 10th N. Y. Vol. Inft.
EDWARD POWERS, Esq., Chicago, Ill.

Doc. No. 40.
From General E. W. Serrell, of New York.

OFFICE OF E. W. SERRELL, CIVIL ENGINEER, }
64 AND 66 BROADWAY, NEW YORK, *Dec.* 9, 1870. }

Sir — I am favored by your letter of the 28th ult., received to-day.

In reply to your request, permit me to say that, from my earliest recollection it was always understood that rain would follow the celebration of the 4th of July in this city, when an unusual display was indulged in, and years ago your suggestion was considered, here, very reasonable, and more than once I have heard of discharges of artillery for the express purpose of bringing on rain.

In the Department of the South, during the war, so well was this thing understood, and the correctness of the theory recognized. that we always looked for rain after heavy cannonading, and at the bombardment of Morris Island, James Island, and several other places, rain followed by sundown, or soon afterwards. I have not my army journal with me, but this I remember well, that both in the South and in Virginia, if rain did not follow a general engagement, we considered it the exception, not the rule; and I think most officers, especially engineers who keep journals, will agree with me, that such is their recorded experience.

Yours obedient servant,
EDWARD W. SERRELL.
EDWARD POWERS, Esq., P. O. Box 45, Chicago, Ill.

Doc. No. 41.

From a Soldier of a Massachusetts Regiment.

RICHMOND, MCHENRY CO., ILL., *Dec.* 26, 1870.

EDWARD POWERS, Esq., Civil Engineer:

Dear Sir — In looking over the columns of a newspaper — *The Watchman and Reflector* — dated October, 1870, I chanced to see an article written over the above signature, making some suggestions, and, at the same time, inviting a statement of facts concerning the effect of the explosion of gunpowder on the atmosphere, in respect to rain, etc., etc. I will scan a three years' experience in the army with a condensation that might be styled "*multum in parvo.*"

I was a soldier in the war of 1861; member of a Massachusetts regiment; was in Gulf Department, and most of the time in Louisiana; was stationed on Ship Island three weeks The troops, 13,000 strong, drilled six hours a day; at least two days in each week, the whole number were put through sham fights, in which some 30,000 rounds of blank cartridges were fired. The day following the first firing was foggy and cloudy; the succeeding night it rained hard. At the second drill of this sort only two brigades fired cartridges, and one battery fired a few rounds. In the night a heavy thunder storm arose, and three men were killed by lightning in one company. Though there was no more rain in the remaining days that we were on the Island, there was much dull weather.

During the bombardment of Forts Jackson and St. Phillip, we had much heavy weather, especially the last days of the bombardment; and, for several days thereafter, rain fell copiously. During, or soon after the various battles and skirmishes in which we were engaged, we almost invariably had heavy weather, and not unfrequently torrents of rain fell.

Up to end of first eighteen months service, though I kept a diary, I had not once thought of the probable cause of sudden changes in the weather, which, according to my diary, had occurred every time that the army moved from one position to another, and, as a consequence, became engaged with the enemy. My attention was first seriously attracted to the matter by repeatedly hearing superstitious soldiers, as

I considered them, remark that fate was against us, because every time we moved we had to wallow in the mud; that we had pleasant weather in camp, but whenever we were set in motion " the rains descended, the floods came," etc., etc. Partly on account of superstitious gossip and gloomy predictions, and partly to gratify my own curiosity, I resolved, from this date to keep a clear and regular account of the weather, both in camp and in the field. I did keep a strict account, but the diary was afterwards burned on the steamer Washington,— so I am obliged to chronicle from memory. The storming of Fort Hudson, May 27, 1863, was followed by torrents of rain. There was much wet weather during the whole siege, extending far inland, and as far south as the Gulf. Immediately after the assault of June 14, there were several days of dull weather, and much rain.

On the Red River campaign, where there was continual fighting for thirty-two days in succession, and several hard contested battles, there was much heavy weather. A heavy thunder storm generally followed the first or second day after a general engagement. This was especially the case where numerous batteries of artillery were brought into action. A severe engagement took place near the Atchafalaya river, La. When the battle began the sun shone clear — not a cloud in sight. Early in the evening the artillery of both contending armies opened a terrific cannonade, which lasted about three hours. Next morning, rain began to fall. The two armies met on the plains of Marksville. The result was a bloody and destructive encounter, followed by nearly a week of rain. This action took place in May, 1864. During my three years' exposure as a soldier. I do not recollect of any considerable engagement not followed by heavy weather and rain. This result was invariably the case when numerous batteries of artillery were brought into action.

I have given a plain statement of facts upon a subject worthy the careful consideration and investigation of scientific men. Looking at this matter in the light of discoveries of the past, that have done so much to enlighten and benefit the human race, it certainly is not impossible, nor even improbable that, at no distant day, the elements may be so controlled that rain shall descend at the will of man.

I am, sir, respectfully, your obedient servant,
MARSHALL M. CLOTHIER.

METEOROLOGICAL OBSERVATIONS.

A few specimens are appended of meteorological observations taken during and immediately subsequent to naval engagements.* It should be stated that the weather was sometimes registered in a different way in the log-books; some of the vessels — especially the gunboats on the Mississippi — having been unsupplied with instruments or with books ruled in the requisite form.

EXPLANATION OF THE SIGNS FOR FORCE OF WIND.

0 Calm.
1 Light Airs.
2 Light Breeze.
3 Gentle Wind.
4 Moderate Wind.
5 Fresh Wind, in which they can just carry top-gallant sails.
6 Strong Wind.
7 Moderate Gale.
8 Fresh Gale.
9 Strong Gale.
10 Heavy Gale.
12 Hurricane.

SIGNS FOR STATE OF THE WEATHER.

b Blue Sky.
c Clouds.
d Drizzling Rain.
f Thick Fog.
g Dark, Stormy Weather.
h Hail.
l Lightning.

m Misty or Hazy.
o Completely overcast.
p Passing Showers.
q Squally.
r Continuous Rain.
t Thunder.
u Ugly and Threatening.

A Star (*) in connection with a letter denotes an extraordinary degree.

*See Doc. No. 35.

152   APPENDIX.

bc denotes Blue Sky with Detached Clouds.
rq   "   Continuous Rain with Squalls of Wind.

The temperature is expressed in degrees Fahrenheit; the barometrical heights in inches.

BOMBARDMENT OF FORT MACON, N. C., APRIL 25, 1862.

Fire opened at 8.40 A. M.

Weather Record of U. S. Steamer Daylight, for April 25, 1862.

Time. A. M.	Wind.		Weath'r	Temperature.	Barometer.
	Direction.	Force.			
1	Calm.	0	b		
2	"	0	b		
3	W. by S.	3	b		
4	S. W.	5	b		
5	"	"	b		
6	"	"	b		
7	"	"	b		
8	"	"	b		
9	"	"	b		
10	"	"	b		
11	"	"	b		
12	"	"	b		
P. M.					
1	S. W. by S.	5	b		
2	"	5	b		
3	"	5	b		
4	S. S. W.	5	b		
5	"	5	b		
6	"	5	c		
7	"	5	c		
8	"	4	c		
9	"	4	c		
10	"	4	c		
11	"	3	c		
12	N. E.	2	c		

APPENDIX. 153

FORT MACON — CONTINUED.

Weather Record of U. S. Steamer Daylight, for April 26, 1862.

TIME. A. M.	WIND. Direction.	Force.	WEATH'R	TEMPERATURE.	BAROMETER.
1	E. N. E.	4	c		
2	"	4	c		
3	"	4	c		
4	N. N. E.	4	c		
5	"	4	c		
6	"	4	c		
7	"	4	c		
8	"	3	c		
9	"	3	c		
10	"	3	c		
11	"	3	c		
12	N. E.	3	c		
P. M.					
1	N. N. E.	5	c		
2	"	5	c		
3	"	5	c		
4	"	5	c		
5	"	5	r		
6	N. E. N.	5	r		
7	"	5	r		
8	N. N. E.	4	c		
9	"	4	c		
10	"	4	c		
11	N. N. W.	5	c		
12	"	5	c		

NOTE.—The foregoing shows rain brought on by firing commenced at a time when there was a fresh wind blowing, as well as a perfectly clear sky overhead. It will be noticed that the wind moved around a little over half the circumference of the horizon between the commencement of the firing and the commencement of the rain.

APPENDIX.

ATTACK ON GENERAL TERRY, ON JAMES ISLAND.

Engagement took place at daybreak, July 16th, 1863, five gunboats assisting in the defence.

Weather Record of the U. S. Steamer New Ironsides, July 16, 1863.

TIME.	WIND.		WEATH'R	TEMPER-	BAROM-
A. M.	Direction.	Force.		ATURE.	ETER.
1	S. W.	1	bc		
2	"	1	bc		
3	"	1	bc	78	29.74
4	W.	1	bc	79	
5	"	1	bc		
6	"	1	bc		
7	"	1	bc		
8	"	1	bc		
9	"	1	bc		
10	"	1	bc		
11	"	1	bc		
12	S. S. W.	3	bc	85	29.74
P. M.					
1	S. S. W.	3	bc		
2	S.	3	bc		
3	"	3	bc		
4	"	3	bc		29.68
5	"	4	c		
6	"	4	c		29.67
7	"	2	c		
8	"	2	c		
9	E.	4	rq		
10	"	4	rq		
11	N. N. E.	2	r		
12	"	2	r		29.71

APPENDIX. 155

BOMBARDMENT OF FORTS SUMTER AND WAGNER, AND THE CUMMINGS POINT BATTERIES, BY THE SIEGE BATTERIES OF GENERAL GILMORE, AUG. 17, 1863.

Fire opened at a very early hour in the morning.

Weather Record of the New Ironsides for Aug. 17, 1863.

TIME. A. M.	WIND.		WEATH'R	TEMPERATURE.	BAROMETER.
	Direction.	Force.			
1	. W.	2	bc		
2	"	2	bc		
3	"	2	bc		
4	"	2	bc	78	29.72
5	"	2	bc		
6	"	2	bc		
7	Variable.	2	bc		
8	"	0 to 1	bc		
9	"	0 to 1	bc		
10	"	0 to 1	bc		
11	"	0 to 1	bc		
12	"	0 to 1	bc		
P. M.					
1	S. E.	2	bc		
2	"	2	c		
3	"	2	c		
4	"	2	c		29.60
5	"	2	c		
6	"	2	r		
7	N. E.	3	r		
8	"	3	r		
9	"	3	r		
10	"	3	c		
11	E.	2	c		
12	"	2	c		29.60

FIRST ATTACK ON FORT FISHER.

Powder Boat Louisiana exploded at a quarter before 2 A. M., Dec. 24, 1864.

Bombardment commenced at half-past 11 A. M., and continued through the day.

Weather Record of Steamer Malvern, Dec. 24, 1864.

Time. A. M.	Wind. Direction.	Wind. Force.	Weath'r	Temperature.	Barometer.
1	W. by S.	2	bc		1.30 A. M. 30.26
2	"	2	bc		1.46 A. M. 30.24
3	W.	3	bc		30.24
4	"	3	bc		30.24
5	W. N. W.	3	bc		
6	"	3	bc	45	30.20
7	W.	4	bc		
8	"	4	bc	40	30.30
9	"	4	bc		30.30
10	W. S. W.	5	bc	42	30.11
11	S. W. by W.	5	bc		30.11
12	S. W.	5	bc	43	30.32
P. M.					
1	S. W.	5	bc	45	
2	"	5	bc		
3	"	4	bc	48	
4	"	3	bc	50	
5	W.	2	bc		
6	"	2	bc		
7	"	1	bc		
8	"	1	bc	45	30.30
9	"	1	bc		
10	"	1	bc	45	30.30
11	"	1	bc		
12	"	1	bc	45	30.30

APPENDIX. **157**

FORT FISHER — CONTINUED.

Bombardment renewed at 7 A. M., and continued for 7 hours.

Weather Record of Steamer Malvern for Dec. 25, 1864.

TIME. A. M.	WIND. Direction.	Force.	WEATH'R	TEMPER- ATURE.	BAROM- ETER.
1	Calm.		bc	52	30.30
2	"		bc	52	30.30
3	"		bc	49	
4	"		bc	49	
5	N. N. E.	1	c	49	30.24
6	North.	1	bc	49	
7	"	1	bc	49	30.30
8	Calm.		— bc		
9	North.	1	bc		
10	"	1	bc	51	30.30
11	N. N. E.	2	bc		
12	"	2	bc	55	30.30
P. M.					
1	N. N. E.	2	c	56	30.26
2	E. N. E.	2	c	54	30.26
3	N. E.	1	c	55	30.22
4	"	1	c	55	30.22
5	"	1	c	55	30.22
6	"	1	c	55	30.10
7	S. E	2	cd	55	
8	"	2	cd		
9	"	2	cd	59	30.10
10	"	3	cd		
11	"	3	cd	55	30.10
12	"				

Fort Fisher — Continued.

Weather Record of Steamer Malvern, for Dec. 26, 1864.

Time. A. M.	Wind. Direction.	Wind. Force.	Weath'r	Temperature.	Barometer.
1					30.00
2	S. E.	3	R	57	29.96
3	"	1	R	57	29.92
4	"	2	o	57	29.90
5	"	4-5	orq	57	29.90
6	S. W.	4-5	orq	57	29.90
7	S. S. W.	4	—		
8	W.	4	orq	60	30.10
9	Baffling.	1	c		
10	"	1	c	66	29.90
11	"	1	c		
12	W. S. W.	2	r	66	29.84
P. M.					
1	Variable.	1	c		
2	"	1	c	59	29.80
3	West.	1	c		
4	"	1	c	56	29.80
5	"	1	bc	58	
6	"	1	bc	50	29,80
7	"	1	bc		
8	"	2	bc	55	29.80
9	Baffling.	1	c		
10	W. by N.	1	c	52	29.79
11	S. S. W.	1	c		
12	West.	1	c	52	29.79

APPENDIX. 159

FIRST DAY'S OPERATIONS AT THE CAPTURE OF FORT FISHER, JAN., 1865.

Heavy bombardment commenced soon after midnight, morning of the 13th. Weather on the 12th, bc, until 8 P. M.; thence until midnight, b.

Weather Record of Flag-ship Malvern, Jan. 13, 1865.

TIME. A. M.	WIND.		WEATH'R	TEMPER- ATURE.	BAROM- ETER.
	Direction.	Force.			
1	S. W. by W.	2	bc	45	30.30
2	"	2	bc		30 30
3	"	2	bc	48	30.30
4	"	2	bc	40	30.29
5	"	2	bc		
6	"	2	bc	40	30.20
7	West.	3	bc		
8	"	3	bc	45	30 20
9	"	6	bc		
10	Calm.	0	bc	46	30.28
11	S. S. W.	2	bc		
12	"	2	bc	57	30.32
P. M.					
1	S. S. W.	2	bc	60	30.30
2	"	2	bc		
3	"	3	H	57	30.35
4	"	4	H	55	
5	"	2	H	55	
6	"	2	H	55	30.22
7	"	2	c		
8	"	2	c	49	30.20
9	"	2	H		
10	"	2	H	50	30.12
11	"	3	H		
12	"	3	H	50	30.08

APPENDIX.

PASSAGE OF THE BATTERIES AT PORT HUDSON, NIGHT OF MARCH 14–15, 1863.

Fire opened at 11.20 P. M., March 14. At midnight, wind from northward, with force 2; weather, bc; barometer, 30.00.

Weather Record of Steam Sloop Hartford, for March 15, 1863.

TIME. A. M.	WIND. Direction.	Force.	WEATH'R	TEMPER- ATURE.	BAROM- ETER.
1	Calm.	0	bc		
2	"	0	bc		
3	"	0	bc		
4	"	0	bc		30.10
5	"	0	bc		
6	"	0	bc		
7	"	0	bc		
8	"	0	bc		
9	"	0	bc		
10	"	0	bc		
11	"	0	bc		
12	"	0	bc		10.30
P. M.					
1	Calm.	0	r		30.10
2	"	0	r*		
3	"	0	r*		
4	"	0	r*		
5	"	0	r*		
6	"	0	r*		
7	"	0	r*		
8	"	0	r*		
9	N. E.	2	r*		
10	"	2	r*		
11	S. S. W.	2	c		
12	"	2	cf		

APPENDIX. 161

FARRAGUT BELOW NEW ORLEANS.

BOMBARDMENT OF FORTS JACKSON AND ST. PHILIP.

More or less firing from 9 A. M. to 6.40 P. M., April 18, 1862.

Weather Record of U. the S. Steam Sloop Hartford, for April 18, 1862.

TIME. A. M.	WIND.		WEATH'R	TEMPER- ATURE.	BAROM- ETER.
	Direction.	Force.			
1	S. E.	5	bc	71	
2	"	"	"	"	30.19
3	"	"	"		
4	"	3	"	"	30.18
5	"	"	"		
6	"	2	"	72	30.20
7	"	"	"		
8	"	"	"	73	30.25
9	"	"	"		
10	"	"	"	79	30.26
11	"	"	"	80	30.25
12	"	"	"		30.16
P. M.					
1	S. E.	2	bc		
2	"	"	"	85	30.23
3	"	"	"		
4	"	"	"		30.22
5	"	"	"		
6	"	"	"		
7	"	"	"	80	
8	"	"	"		
9	"	"	"	75	30.24
10	"	"	"	75	"
11	"	"	"		
12	"	"	"		30.21

BOMBARDMENT OF FORTS JACKSON AND ST. PHILIP—CONTINUED.

More or less firing from 6.40 A. M. to midnight, April 19.

Weather Record of the Hartford, for April 19, 1862.

TIME. A. M.	WIND. Direction.	Force.	WEATH'R	TEMPER- ATURE.	BAROM- ETER.
1	S. E.	1	bc		
2	"	1	bc		
3	"	1	bc		
4	"	1	bc	72	30.20
5	"	1	bc		
6	"	1	bc	70	30.20
7	"	1	bc		
8	"	1	bc		
9	"	1	bc	76	30.19
10	"	1	bc		
11	"	1	bc		
12	"	1	bc	84	30.15
P. M.					
1	S. S. W.	1	bc		
2	"	1	bc		
3	"	1	bc		
4	"	1	bc	83	30.10
5	"	1	bc		
6	S. S. E.	1	bc	78	30.18
7	"	1	bc		
8	"	1	bc	80	30.08
9	South.	1	b	75	
10	"	1	b		
11	"	1	b		
12	"	2	b	73	

APPENDIX. 163

BOMBARDMENT OF FORTS JACKSON AND ST. PHILIP — CONTINUED.

Weather Record of the Hartford, for April 20, 1862.

TIME. A. M.	WIND.		WEATH'R	TEMPER- ATURE.	BAROM- ETER.
	Direction.	Force.			
1	S. W.	2	bc	74	
2	"	2	bc		30.05
3	"	2	bc		
4	"	2	bc		29.96
5	"	2	bc		
6	"	2	rpc	70	29.96
7	W. S. W.	2	rpc		
8	"	2	rpc		
9	"	2	rpc		
10	"	2			
11	"	2			
12	"	2		63	29.98
P. M.					
1	W. S. W.	3	c		
2	"	9	c	62	29.98
3	"	4	c	60	
4	"	9	c	58	
5	N. W.	5	c	56	
6	"	5	c	55	30.07
7	"	5	c		
8	"	5	c		
9	N. W. by W.	5	bc	56	30.00
10	"	4	bc		30.00
11	"	3	bc		30.00
12	"	3	bc		30.00

FARRAGUT BELOW NEW ORLEANS — Continued.

Grand Attack on Forts Jackson and St. Philip — Passage of the Forts, and Destruction of the Rebel Fleet.

Fight commenced before daylight, April 24, 1862.

Weather Record of U. S. Steam Sloop Hartford, for April 24, 1862.

Time. A. M.	Wind. Direction.	Force.	Weath'r	Temperature.	Barometer.
1	S. by W.	1	b	75	30.20
2	"	1	b	75	30.20
3	"	1	b	75	30.20
4	"	1	b	75	30.20
5	"	1	b	75	30.20
6	"	1	b	75	30.20
7	"	2	b	75	30.20
8	S. W.	2	b	70	30.20
9	"	2	b	70	30.20
10	"	2	b	73	30.20
11	"	2	b	"	
12	"	2	b	75	30.21
P. M.					
1	S. W.	2	b		
2	"	2	b		
3	"	2	b		
4	"	2	b	78	30.21
5		2	b		
6	E. S. E.	2	bc	78	
7	"	2	bc		30.21
8	"	2	bc		
9	"	2	bc		
10	"	2	bc		
11	"	2	bc		30.15
12	"	2	bc		

APPENDIX. 165

FORTS JACKSON AND ST. PHILIP — CONTINUED.

DAY AFTER THE BATTLE.

Weather Record of Steam Sloop Hartford, for April 25, 1862.

TIME. A. M.	WIND.		WEATH'R	TEMPER- ATURE.	BAROM- ETER.
	Direction.	Force.			
1	Calm.	0	b		
2	"	0	b		
3	"	0	b		
4	"	0	b		30.10
5	"	0	b		
6	"	0	b		30.10
7	"	0	b		
8	"	0	b	30.15	30.15
9	"	0	b		
10	Variable.	0	o		
11	"	0	o		
12	S. S. W.	2	r		
P. M.					
1	S. S. W.	2	r		
2	"	2	r		
3	"	2	r		
4	"	2	r		30.04
5	"	2	c		
6	"	2	c		
7	"	2	c		
8	"	2	c		
9	N. W.	3	bc		
10	"	3	ol		
11	"	3	ol		
12	"	3	bc		30.88

FARRAGUT AT THE ENTRANCE OF MOBILE BAY.

PASSAGE OF THE FORTS, NAVAL BATTLE, AND CAPTURE OF THE "TENNESSEE," AUG. 5, 1864.

Engagement commenced at about 6.45 A. M. Shower in afternoon of day before.

Weather Record of U. S. Steam Sloop Hartford, for A. M. Aug. 5, 1864.

TIME. A. M.	WIND.		WEATH'R	TEMPERATURE.	BAROMETER
	Direction.	Force.			
1	S. E.	1	bc		
2	"	1	bc	81	30.04
3	S. S. E	1	bc		
4	"	1	bc	80	30.04
5	"	1	bc		
6	S. W.	1	bc	80	
7	"	1	bc		
8	"	2	bc	82	
9	"	1	bcr		
10	"	1	bcr	82	
11	"	1	bc		
12	"	1	bc	83	

APPENDIX. 167

FARRAGUT AT ENTRANCE OF MOBILE BAY — Continued.

Shelling of Fort Gaines by the "Chickasaw."

Weather Record of the Hartford, for August 6, 1864.

Time. A. M.	Wind. Direction.	Force.	Weath'r	Temperature.	Barometer.
1	S. E.	1	bc	82	
2	"	1	bc	82	
3	"	1	bc	82	
4	"	1	bc	82	
5	"	1	bc		
6	"	1	bc		
7	"	1	bc		
8	"	1	bc		
9	"	1	bc	87	30.29
10	"	1	bc	87	30.27
11	"	1	bc	87	30.27
12	"	1	bc	87	30.27
P. M.					
1	S. S. E.	1	bc	82	30.29
2	"	1	bc	82	30.29
3	"	1	bc	82	30.29
4	"	1	bc	82	30.29
5	"	1	bc		30.29
6	"	1	bc	82	30.29
7	"	1	bc	81	30.28
8	"	1	bc	79	30.28
9	"	1	bc	79	30.27
10	"	1	bc	79	30.27
11	"	1	bc	79	30.27
12	"	1	bc	79	30.27

FARRAGUT AT ENTRANCE OF MOBILE BAY — Continued.

Weather Record of the Hartford, for August 7, 1864.

Time. A. M.	Wind. Direction.	Force.	Weath'r	Temperature.	Barometer.
1	S. by W.	1	bc		
2	"	2	bc		
3	Calm.	0	bc		
4	"	0	bc		
5					
6					
7					
8					
9	S. by E.	1	bc	89	30.29
10	"	1	bc		
11	"	1	bc	88	30.30
12	"	1	bc		

P. M.					
1	S. by E.	1	bc		
2	"	1	bc	88	30.30
3	"	1	bc		
4	"	1	bc	88	30.30
5	"	1	bc		
6	E. S. E.	1	bc	88	30.30
7	"	1	bc	84	
8	"	1	bc		30.30
9	"	1	bc	83	
10	E. by S.	1	oc		30.26
11	"	1	oc	83	
12	"	1	oc	83	30.26

APPENDIX. 169

FARRAGUT AT ENTRANCE OF MOBILE BAY—Continued.

Weather Record of the Hartford, for August 8, 1864.

Time.	Wind.		Weath'r	Temper-	Barom-
A. M.	Direction.	Force.		ature.	eter.
1	E.	1	bc		
2	"	1	bc	82	30.30
3	"	1	bc		
4	"	1	bc	82	30.30
5	"	1	bc		
6	N. N. W.	1	bc	82	30.30
7	"	1	bc		
8	N. by E.	1	bc	84	30.34
9	"	1	bc		
10	"	1		84	30.30
11	N.	1			
12	"	1		87	30.28
P. M.					
1	N. W.	2	bc		
2	W.	2	bc	89	30.28
3	"	2	bc		
4	"	2	bc	89	30.28
5	"	2	bc		
6	"	2	bc	89	30.28
7	N. E.	1	qr		
8	N. N. E.	1	qr	85	30.27
9	Calm.	0	ol		
10	"	0	ol	83	30.21
11	S. W.	1	bcl	82	
12	"	1	bc	82	30.21

FARRAGUT AT ENTRANCE OF MOBILE BAY.—Continued.

Weather Record of the Hartford, for A. M., Aug. 9, 1864.

Time. A. M.	Wind. Direction.	Wind. Force.	Weath'r	Temperature.	Barometer.
1	N. E.	1	bcl		
2	"	1	bcl	81	30.15
3	N.	1	bcl		
4	N. E. by N.	1	bcl	81	30.15
5	"	1	bcl		
6	Calm.	0	bcl	81	30.19
7	"	0	bc		
8	"	1	bc	81	30.19
9	E.	1	bc		
10	"	1	bc	81	30.19
11	"	1	bc		
12	"	1	bc	81	30.19

Shelling of Fort Morgan by Several Vessels of the Fleet.

Action commenced at noon, Aug. 9, and ended at 4 P. M.

Weather Record of the Hartford, for P. M., Aug. 9, 1864.

Time. P M.	Wind. Direction.	Wind. Force.	Weath'r	Temperature.	Barometer.
1	E.	1	bc		
2	S.	1	bc	84	30.33
3	"	2	oc		
4	"	2	oc	83	30.11
5	"	2	oc		
6	"	2	oc	82	30.11
7	"	2	oc		
8	"	2	oc	82	30.11
9	"	2	bc		
10	"	1	bc	81	30.24
11	"	1	bc		
12	"	1	bc	81	30.24

APPENDIX. 171

DAY AFTER THE SHELLING OF FORT MORGAN.

Weather Record of the Hartford, for Aug. 10, 1864.

TIME. A. M.	WIND.		WEATH'R	TEMPER- ATURE.	BAROM- ETER.
	Direction.	Force.			
1	S. E.	2	gl		
2	"	3	gl	80	30.16
3	S. by W.	2	gl		
4	"	1	gl	81	30.14
5	Baffling.	1	gl		
6	"	1	gl	79	30.14
7	"	1	gl		
8	"	1	gl	81	30.16
9	"	1	gl		
10	S. W.	1	gr	80	30.14
11	"	4	gr		
12	S.	3	gr	80	30.12
P. M.					
1	S.	2	o		
2	S. E.	1	o	76	30.15
3	"	1	o		
4	"	1	od	77	30.10
5	"	1	o		
6	"	1	o	76	30.10
7	"	2	bdt		
8	"	2	bdt	78	30.07
9	"	2	bcl		
10	"	2	bcl	78	30.07
11	S. S. E.	3	bcl		
12	"	3	lg	78	30.06

The following remarks are also entered in the Log: "From midnight to 4 A. M. had a squall of rain from the east. At 9.30 A. M. a heavy rain squall from the southwest."

www.ingramcontent.com/pod-product-compliance
Lightning Source LLC
Chambersburg PA
CBHW031453160426
43195CB00010BB/966